Chemoreceptors and Ventilation: The Link Between Blood Gases and Breathing

Georgia

TABLE OF CONTENTS

CHAPTER I: INTRODUCTION

The monitoring of oxygen (O_2), carbon dioxide (CO_2) and proton (H^+) levels in arterial blood is essential for the control of respiration (Prabhakar, 2000; Teppema and Dahan, 2010; Guyenet and Bayliss, 2015; Guyenet et al., 2019). Within seconds, arterial hypoxia (low pO_2), hypercapnia (high pCO_2) and/or acidosis (high H^+ levels) stimulate breathing, and this ventilatory adjustment depends critically on the chemosensory abilities of the carotid bodies and respiratory control brainstem nuclei (Prabhakar, 2000), and the capability of the sympathetic nervous system to modulate the activities of these respiratory-related structures. In this introductory book chapter, we will discuss the structures and functions of the main peripheral chemoreceptor, the carotid body, and the numerous central chemoreceptors. In addition, we will reveal proposed mechanisms behind the ventilatory responses to hypoxia, hypercapnia, and hypoxia-hypercapnia gas exposure, and how sympathetic innervation to peripheral and central chemosensory structures regulates their activity in response to changes in blood gas chemistry.

THE CAROTID BODY

The carotid body (CB) is the main peripheral arterial chemoreceptor that rapidly responds to changes in blood pO_2, pCO_2, and pH by regulating breathing (Prabhakar, 2000; Atanasova et al., 2011; Ortega-Sáenz and López-Barneo, 2020). It is a small bilateral organ strategically located in the bifurcation region of the common carotid artery (Atanasova et al., 2011; Ortega-Sáenz et al., 2020) (**Figure 1a**), and is derived from neural crest progenitor cells of sympathoadrenal lineage, which migrate during embryogenesis

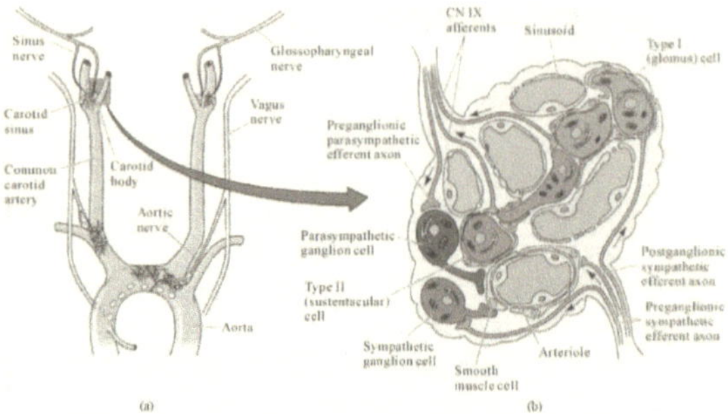

Figure 1. Location and interactions between cellular elements in the carotid body (Boron and Boulpaep, 2012). (a) Location of carotid body in carotid bifurcation. (b) Schematic representation of the microscopic anatomy of the carotid body.

from the superior cervical ganglion (SCG) to the carotid bifurcation (Ortega-Sáenz and López-Barneo, 2020). It is highly vascularized, with numerous fenestrated capillaries, and receives blood supply through the occipital artery, a branch arising from the external carotid artery (Prabhakar, 2000).

STRUCTURE OF THE CAROTID BODY GLOMERULI

The CB parenchyma consists of clusters of chemosensitive, neuron-like glomus cells (also called type I or chief cells) (i.e., glomeruli) that are innervated by sensory afferent fibers of the carotid sinus nerve (CSN). Glomus cells are ovoid shaped and contain numerous clear and dense-core secretory vesicles filled with dopamine and numerous other neurotransmitters, which allows antibodies against tyrosine hydroxylase (TH) to be used as suitable cell markers (Ortega-Sáenz and López-Barneo, 2020). Sympathetic and parasympathetic neuronal bodies and efferent fibers terminating on glomus cells and blood vessels are also located within the carotid body (Atanasova et al., 2011; Ortega-Sáenz and López-Barneo, 2020). Glomus cells are encompassed by interdigitating processes of glia-like type II or sustentacular cells, which stain with the cell marker glial fibrillary acidic protein (GFAP) (Atanasova et al., 2011; Ortega-Sáenz and López-Barneo, 2020) (**Figure 1b**). Early researchers have postulated the existence of glomus cell subpopulations, which may serve as progenitor glomus cells (Hellström, 1975; McDonald and Mitchell, 1975), however recent studies have identified pre-differentiated CB neuroblasts located at the periphery of a CB glomeruli that can undergo mitotic divisions before final maturation, unlike mature CB glomus cells, which are postmitotic (Sobrino et al., 2018). In addition, research has shown that type II cells, or a subpopulation of them, have been characterized as multipotent stem cells that support CB growth when exposed to stressful conditions, such as chronic hypoxia (Pardal et al., 2007).

CAROTID BODY GLOMUS CELL CHEMOSENSORY TRANSDUCTION

CB glomus cells form synapses with neighboring glomus cells and chemosensory afferent fibers of the CSN. They behave as presynaptic-like neurosecretory elements, containing a plethora of neurotransmitters and neuromodulators (e.g. ATP, dopamine, adenosine, GABA, serotonin, acetylcholine, opioids, histamine, substance P, endothelin-1, vasoactive intestinal peptide, and angiotensin II, etc.) that are stored primarily in synaptic vesicles and released by calcium-dependent exocytosis (Ureña et al., 1994; Pardal et al., 2000; Buttigieg and Nurse, 2004). ATP is the main excitatory neurotransmitter acting on postsynaptic ionotropic (P2X2/3) purinergic receptors located on CSN terminals (Alcayaga et al., 2000; Zhang et al., 2000; Rong et al., 2003). In addition, ATP exerts inhibitory action when binding to metabotropic P2Y receptors on glomus cells to block hypoxia-induced cell depolarization (Xu et al., 2005; Tse et al., 2012). Other neurotransmitters, such as dopamine, exerts a strong autocrine inhibitory action on glomus cells (Benot and López-Barneo, 1990), as well as its paracrine inhibitory effects on nerve endings of the CSN (Zhang et al., 2018). In addition to glomus cells' chemically mediated auto- and/or paracrine effects, they can be electrically interconnected by gap junctions, although the functional significance of these is not well known (Ortega-Sáenz and López-Barneo, 2020). Recent studies have also published interesting data revealing that type II cells play a critical role in chemotransduction. Electron microscopy analyses have shown that the vast majority of secretory vesicles in glomus cells lay facing type II cell membranes and are not clustered in front of postsynaptic nerve endings (Platero-Luengo et al., 2014). Thus, glomus and type II cells form chemical synapses that are unique from glomus cell-nerve ending chemosensory synapses, suggesting that the glomus cell–nerve ending synapse is indeed a tripartite synapse (Nurse, 2014; Tse et al., 2012).

CAROTID BODY PLASTICITY

The CB is extremely plastic, which is crucial for physiological adaptation to altering environments and pathological conditions. For instance, living at high altitude requires an increased sensitivity of the CB to the reduction in O_2 tension in the air, therefore resulting in a constant hyperventilatory response to prevent a deleterious decrease in arterial O_2 levels (Joseph and Pequignot, 2009; Bisgard, 2000). Plasticity of the CB depends on (1) functional changes occurring in the expression of ion channels and/or the availability of peptides and neurotransmitters that regulate cellular interactions within the CB chemosensitive glomus cells (Powell, 2007), and (2) structural changes occurring in the ability of the CB to increase in size several fold and thus increase afferent inputs to the central respiratory control centers in the brainstem (Joseph and Pequignot, 2009).

Studies have shown that chronic sustained hypoxia (CSH) upregulates the expression of Na^+ and/or Ca^{2+} channels and dampens K^+ current amplitude (Powell, 2007), therefore resulting in enhanced CB glomus cell excitability. In addition, hypoxia does not alter the levels of TASK1 channels expressed in cultured neonatal glomus cells, however, hypoxic inhibition of these background K^+ currents are augmented after 48h of CSH in these cells (Ortiz et al., 2009). Moreover, CSH (1) strongly induces nitric oxide (NO) synthase, as well as TH, the rate-limiting enzyme in the synthesis of dopamine, (2) downregulates the expression of D2 dopamine receptors; and (3) upregulates nicotinic cholinergic receptors in CB postsynaptic chemoafferent fibers, thus increasing the excitatory effect of acetylcholine (Bisgard, 2000; Powell, 2007). CB functional plasticity is also activated by chronic intermittent hypoxia (CIH), a common phenomenon in humans that is associated with recurrent obstructive apneas occurring during sleep (i.e., sleep apnea) (Ortega-Sáenz and López-Barneo, 2020). CIH increases the hypoxic ventilatory response (HVR) due to long-term potentiation of the CB chemoreflex, a form of plasticity resulting from increased CB excitation during acute episodes of hypoxia (Peng et al., 2003; Rey et al., 2004; Marcus et al., 2010). Furthermore, inhibition of TASK-like K^+ channel conductance in CB glomus cells is increased in animals exposed to CIH (Iturriaga et al., 2014), and CIH-induced modifications in CB glomus cells has been proposed to involve a mitochondrial reactive oxygen species (ROS)-mediated imbalance in the hypoxia inducible factor (HIF)-1 and HIF-2 signaling pathways (Yuan et al., 2013).

In addition to the functional changes described above, CSH causes dramatic structural changes in the CB, including a two- to threefold increase in size due to angiogenesis and the generation of new glomus cell clusters (Ortega-Sáenz and López-Barneo, 2020). For example, numerous studies suggest that some chemical molecules released from glomus cells (e.g., endothelin-1, ATP, or acetylcholine), may play crucial roles in the activation of quiescent multipotent CB stem cells (with a type II phenotype) or CB neuroblasts, and induce their proliferation and differentiation into mature glomus, endothelial and smooth muscle cells (Pardel et al., 2007; Annese et al., 2017; Sobrino et al., 2018; Platero-Luengo et al., 2014; Macías et al., 2014). CB stem (type II) cells and predifferentiated CB neuroblasts expressing voltage-dependent ion channels and catecholaminergic markers (TH+), are not intrinsically sensitive to hypoxia and thus their activation relies on chemical synapses they form with the mature O_2-sensitive glomus cells (Sobrino et al., 2018; Platero-Luengo et al., 2014). Mature O_2-sensitive glomus cells release endothelin-1 during hypoxia that bind to endothelin-1 receptors present on stem cells thus inducing proliferation (Platero-Luengo et al., 2014). Contrarily, downregulation of HIF2α in adult mice results in an inhibition of CB cell proliferation during CSH (Hodson et al., 2016). Unlike mature glomus cells, CB neuroblasts can undergo mitotic divisions before their final maturation, which is triggered by hypoxia as well as secreted neurotransmitters (e.g., acetylcholine and ATP), released from glomus cells (Ortega-Sáenz and López-Barneo, 2020). This phenotypic change in the CB TH+ cell population (stem cells differentiating from non-O_2-sensitive to O_2-sensitive) may have a significant pathogenic role in CB oversensitivity observed in conditions, such as sleep apnea (Ortega-Sáenz and López-Barneo, 2020).

CAROTID BODY DYSFUNCTION AND DISEASES

Dysfunction in the CB affects the susceptibility to respiratory depression in addition to the pathogenesis of highly prevalent diseases (e.g., hypertension, chronic heart failure or sleep apnea) (Ortega-Sáenz and López-Barneo, 2020). As discussed above, the CB rapidly responds to decreases in arterial pO_2, therefore inhibition of the CB may have fatal consequences. Evidence has shown that maturation of the CB chemoreflex is critical for adaptation of newborn mammals to extrauterine life, and developmental defects of the CB *in utero* that result in CB inhibition, causes respiratory disorders later in life, such as central congenital hyperventilation or sudden infant death syndromes (López-Barneo et al., 2016; Weir et al., 2005). Even so, the most recurrent cases of CB inhibition happen after surgical removal of the CB or the administration of drugs. For instance, patients with bilateral CB resection display a practically abolished HVR (Teppema and Dahan, 2010; Timmers et al., 2003), even though they seem to function unaffected in a normoxic environment. CB inhibition is also a side effect of numerous sedatives, anesthetics, and analgesic drugs (Ortega-Sáenz and López-Barneo, 2020). For instance, intracarotid administration of methionine-enkephalin and morphine, activate K^+ channels and inhibit Ca^{2+} channels, consequently interfering with glomus cell neurosecretory response (López-Barneo et al., 2016), and therefore depressing overall CB chemosensory afferent discharges in the CSN (Kirby and McQueen, 1986; McQueen and Ribeiro, 1980).

Along with inhibition of the CB, overactivation of the CB is also a CB dysfunction that leaves patients highly susceptible to prevalent diseases (Ortega-Sáenz and López-Barneo, 2020). CB overactivation is recognized as a major cause of the exaggerated sympathetic outflow characteristic of disorders such as chronic heart failure, hypertension, and sleep apnea (Ortega-Sáenz and López-Barneo, 2020; López-Barneo et al., 2016). Even though the mechanistic pathways behind CB overexcitability remain poorly understood, evidence has shown that bilateral CB ablation results in a reduced sympathetic tone and thus reversal of some of the related cardiovascular alterations, such as high blood pressure (Paton et al., 2013; Del Río et al., 2013, 2016). Nonetheless, the clinical translation of surgical CB ablations must be done cautiously since in a pilot study, bilateral CB ablation improved the sympathetic imbalance in chronic heart failure patients, but also augmented the occurrence of nighttime hypoxia, specifically in patients with concomitant sleep apnea (Niewinski et al., 2017). Therefore, the development of drug therapies to selectively modulate CB chemosensory activity is an ideal alternative to surgery and can increase the translation of CB research to a clinical reality (Ortega-Sáenz and López-Barneo, 2020).

THE CAROTID SINUS NERVE

The carotid sinus nerve (CSN) branches from the glossopharyngeal nerve (Cranial Nerve (CN) IX) with cell bodies localized in the petrosal ganglion (PG). It communicates with the sympathetic trunk (usually at the level of the superior cervical ganglion) and the vagal nerve (main trunk, pharyngeal branches, or superior laryngeal nerve) (Porzionato et al.,

2019). The CSN travels on the anterior aspect of the internal carotid artery to reach the carotid sinus, CB, and/or intercarotid plexus (Porzionato et al., 2019). In the carotid sinus, type I (dynamic) carotid baroreceptors are innervated by large, myelinated A-fibers of the CSN, whereas type II (tonic) baroreceptors are innervated by small, A- and unmyelinated C-fibers of the CSN (Porzionato et al., 2019). In the CB, chemosensory afferent fibers of the CSN are primarily activated by acetylcholine and ATP, released from glomus cells in response to decreases in pO_2 and pH, and/or elevations in pCO_2 (Porzionato et al., 2019; Gonzalez et al., 1994; Eyzaguirre and Zapata, 1984; Kikuta et al., 2019; Peers and Buckler, 1995; Prabhakar, 2000; Tse et al., 2012). Once stimulated, the CSN fibers course through the solitary tract (ST) and synapse on neurons in different regions of the NTS. For instance, chemosensory afferents of the CSN project to the cNTS, and baroreceptor inputs project to the cNTS, medial (m)NTS, dorsomedial (dm)NTS, and dorsolateral (dl)NTS (Porzionato et al., 2019). The chemosensory component of the CSN elicits sympatho-activation, and the baroreceptor component stimulates sympatho-inhibition (Porzionato et al., 2019), suggesting that in pathologies, such as refractory hypertension and heart failure (characterized by increased sympathetic activity), baroreceptor electrical stimulation, and CB removal may be a potential therapy.

CAROTID SINUS NERVE INNERVATION OF THE CAROTID BODY

This section will focus on CSN innervation of the CB. Innervation of the carotid sinus and intercarotid plexus will not be discussed further but can be read about in a fantastic review by Porzionato et al. (2019). As mentioned above, CB glomus cells are innervated by chemosensory afferent fibers of the CSN, a branch of CN IX, with cell bodies localized in the caudal region of the petrosal ganglion (Iturriaga and Alcayaga, 2004; Prabhakar and Overholt, 2000; Ortega-Sáenz and López-Barneo, 2020). These CSN afferent fibers show immunoreactivity for the rate-limiting enzyme for catecholamine synthesis, TH (Massari et al., 1996). Decreases in pO_2 and pH, and/or elevations in pCO_2 trigger CB chemosensitive glomus cells to release neurotransmitters (e.g. ATP and acetylcholine) that excite closely apposed CSN terminals, causing increased firing of these afferents (Gonzalez et al., 1994; Eyzaguirre and Zapata, 1984; Kikuta et al., 2019; Peers and Buckler, 1995; Prabhakar, 2000; Tse et al., 2012). CSN chemoafferents then relay this signal to the cNTS within the brainstem (Song et al., 2011), and the cNTS then stimulates the rostral ventrolateral medulla, which in turn projects to and stimulates the intermediolateral nucleus of the spinal cord (Porzionato et al., 2019). Overall, increased firing from these afferent fibers triggers downstream responses to restore arterial blood gas status and hemodynamic homeostasis (Nurse, 2005; Nurse, 2010; Prabhakar, 2000).

Studies using ultrastructure have reported (1) reciprocal synapses between CSN chemoafferent fibers and chemosensitive glomus cells (McDonald and Mitchell, 1975), and (2) postnatal developmental changes in CB innervation, including a decrease in vesicle-rich nerve endings and an increase in mitochondria-rich ones (interpreted as a decrease in efferent terminations and an increase in afferent fibers) (Bollé et al., 2000), in addition to proliferation of Schwann cells that line CSN fibers (Wang and Bisgard, 2005). Interestingly, evidence has shown that a major role in the production of these developmental changes is played by trophic factors, such as glial cell line-derived neurotrophic factor (GDNF) and brain-derived neurotrophic factor (BDNF), acting on CB cell populations in autocrine or paracrine ways (Porzionato et al., 2008). Studies have also shown that CB nerves exposed to environmental stimuli, such as chronic hypoxia, exhibit a decrease in the densities of calcitonin gene-related peptide (CGRP) and substance P fibers and an increase in vasoactive intestinal peptide (VIP) fiber density (Poncet et al., 1994; Kusakabe et al., 1998, 2003). This effect is reversed a few weeks after termination of hypoxia and return to a normoxic state (Kusakabe et al., 2004). Conversely, postnatal exposure of a rat to four weeks hyperoxia resulted in a 41% decrease in unmyelinated sensory nerve fibers in the CSN, together with a 32% reduction of the number of TH-positive neurons in the petrosal ganglion, thus suggesting hyperoxia-induced CB hypoplasia (Erickson et al., 1998). In addition to the numerous afferent fibers present in the CSN, researchers have revealed the existence of some efferent fibers as well (Porzionato et al., 2019). For instance, respiratory-modulated glossopharyngeal efferents, focused on the carotid sinus or CB, are present in the CSN and are thought to represent a regulatory mechanism modulating the sensitivity of chemo- and/or baroreceptors (Kostreva et al., 1984; Nurse and Piskuric, 2013).

CAROTID SINUS NERVE AS A THERAPEUTIC TARGET

Stimulation of the CSN has been proposed for treatment therapy of heart failure (Gronda et al., 2014), drug-resistant hypertension (Neistadt and Schwartz, 1967), and epilepsy (Patwardhan et al., 2002; Tubbs et al., 2002). With regards to treatment for epilepsy, an experimental study in a canine model showed that stimulation of the glossopharyngeal nerve, through the CSN, produced better results for seizure control (i.e., CSN stimulation resulted in no cardiac side-effects) compared to vagal nerve stimulation (Patwardhan et al., 2002). Moreover, hypertension and heart failure show excessive sympathetic nerve activity that may be related to hyperactive CB chemoreflex activity (Paton et al., 2013). For instance, authors have reported that a decrease in the density of VIP fibers in spontaneously hypertensive rats resulted in an increased chemoreceptor sensitivity, since it is known that VIP has an inhibitory effect on CB chemoreception (Takahashi et al., 2011). In addition, a study by Narkiewicz et al. (2016) showed that unilateral CB removal resulted in significant reductions in blood pressure and muscle sympathetic nerve activity in a subpopulation of patients affected by drug-resistant hypertension.

CENTRAL CHEMORECEPTORS

Central chemoreception monitors and integrates information on (1) brain blood flow and metabolism, (2) alveolar ventilation, and (3) acid-base balance, and, in response, can alter breathing, airway resistance, arousal, and blood pressure (sympathetic tone) (Nattie and Li, 2012). Central chemoreceptors detect increases in (1) arterial pCO_2 and (2) protons (H^+) produced via the carbonic anhydrase-catalyzed reaction of CO_2 and H_2O from the increased availability of blood CO_2, within the brain (Nattie and Li, 2012). Functionally, central chemoreceptors, along with the carotid bodies, regulate (1) arterial pCO_2, and (2) blood and body pH in response to acid-base disturbances (Nattie and Li, 2012). Moreover, central chemoreceptors provide a tonic excitation at the physiological (i.e., normal) arterial pCO_2 level in order to sustain functional connectivity between brainstem respiratory neurons responsible for generating eupneic breathing (Nattie and Li, 2012). Overall, central chemoreceptors respond to small fluctuations in blood pCO_2 to regulate normal gas exchange, and to big changes in pCO$_2$ to reduce acid-base alterations (Nattie and Li, 2012).

LOCATION OF CENTRAL CHEMORECEPTORS

Central chemoreceptors were originally localized to areas on the ventral surface of the medulla, by experiments that used direct application of acidic fluids in anesthetized animals (Nattie and Li, 2012). However, current substantial evidence suggests the presence of a widespread distribution of central chemoreceptors in many locations within the brainstem, cerebellum, hypothalamus, and midbrain (Lamb, 1966; Li et al., 2006;

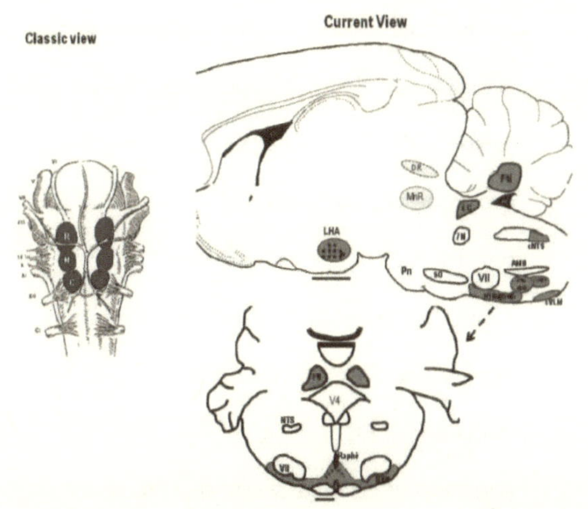

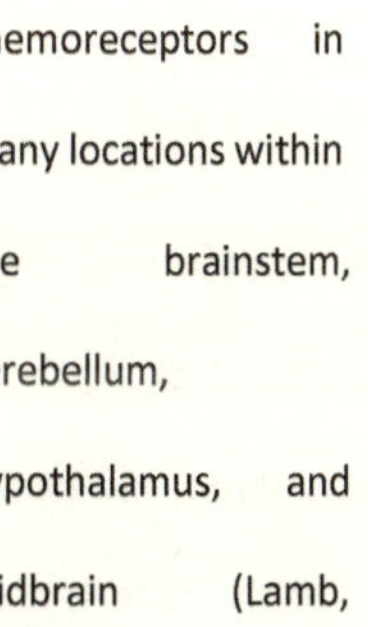

Figure 2. Schematic location of central chemoreceptor sites (taken from figure 3 in Nattie and Li, 2012). Locations of central chemoreceptors; the classic view: chemoreception located at ventral medullary surface (left panel) and the current view: chemoreception is widely distributed in hindbrain (right panel). Abbreviations: R, rostral; M, middle; C, caudal; LHA, lateral hypothalamus; DR, dorsal raphe; FN, fastigial nucleus; 4v, fourth ventricle; LC, locus ceruleus; 7N, facial nerve; cNTS, caudal nucleus tractus solitarious; AMB, ambiguous; VII, facial nucleus; SO, superior olive; PBC, pre-BÖotzinger Complex; rVRG, rostral ventral respiratory group; cVLM, caudal ventrolateral medulla; RTH/pFRG, retrotrapezoid nucleus/parafacial

Linton et al., 1976; Messier et al., 2004; Nattie, 1998; Thannickal et al., 2000; Schlaefke et al., 1979). These locations include (**Figure 2**): (1) the retrotrapzoid nucleus (RTN) (Li and Nattie, 2002, 2008), (2) the caudal NTS (Nattie and Li, 1994) (3) the rostral medullary raphe (MR) (raphe magnus) (Hilaire et al., 2004; Nattie and Li, 1996), (4) the caudal MR (Hilaire et al., 2004; Hitzig et al., 1994) (raphe obscurus) indirectly via the RTN (Deng et al., 2007), (5) the region just dorsal to the caudal ventral medullary surface (Cummings et al., 2010), (6) the fastigial nucleus of the cerebellum (Marjanovic et al., 1998), (7) the pre-Bötzinger region (Kogo and Arita, 1990), and (8) the orexin neuron containing regions of the hypothalamus (Li N, Li A, Nattie, E; personal communication). Despite these numerous individual central chemoreception sites, there is a well-recognized understanding among authors in this field that central chemoreception involves a system in which the sites are interconnected and display varying levels of effectiveness dependent upon gender, stimulus intensity, and arousal state (Nattie and Li, 2012). In addition, investigators support the idea that the ventilatory response to central chemoreceptor activation is extremely dependent on the simultaneous level of peripheral chemoreceptor activation (Nattie and Li, 2012).

FUNCTION OF CENTRAL CHEMORECEPTORS

CO_2 and H^+ sensitive central chemoreceptors continuously monitor the levels of CO_2 in arterial blood and in alveolar gas as such to provide rapid feedback to the brainstem respiratory control centers in order to maintain blood gas status and hemodynamic homeostasis (Nattie and Li, 2012; Nurse, 2005; Nurse, 2010; Prabhakar, 2000). As mentioned earlier, central chemoreceptors are associated with two primary physiological functions: (1) the preservation of a constant, normal arterial pCO_2 by using a negative feedback control loop, and (2) the preservation of a constant pH via ventilatory exchange of CO_2 at the alveoli to minimize metabolically produced acid-base disturbances (Nattie and Li, 2012). These functions of central chemoreceptors are largely dependent upon peripheral chemoreceptors. In order words, central and peripheral chemoreceptors are interdependent and work together to respond quickly and sensitively to changes in blood pCO_2, thus allowing rapid regulation of alveolar respiration relative to metabolism (Nattie and Li, 2012). Overall, central chemoreceptors are responsive to the pH of brain interstitial fluid, which allows a widespread, integrative neuronal network of chemosensitive, respiratory-related circuits to sense and respond to alterations in (1) cerebral blood flow, (2) alveolar ventilation, and (3) cerebral metabolism (Nattie and Li, 2012).

VENTILATORY RESPONSES TO HYPOXIA, HYPERCAPNIA, AND HYPOXIA-HYPERCAPNIA

The ventilatory responses to hypoxic (HX), hypercapnic (HC), and hypoxic-hypercapnic (HH) gas exposures are highly dependent on the O_2 and CO_2 sensing abilities of the main peripheral chemoreceptors (i.e., the carotid bodies) and the central chemoreceptors described above. The central and peripheral chemoreceptors depend on each other to respond quickly and efficiently to changes in blood pO_2 and pCO_2 levels and restore arterial blood gas homeostasis.

HYPOXIC VENTILATORY RESPONSE

Since the CB is well recognized as the major O_2 sensor, it is essential for immediate survival under hypoxic conditions (Ortega-Sáenz and López-Barneo, 2020). During mammalian fetal life, the CB does not contribute significantly to breathing, however, in the early weeks after birth, the CB gradually develops hypoxic chemosensitivity and the slope of the hypoxic stimulus response curve increases (Porzionato et al., 2019). In neonates, the HX ventilatory response appears as a biphasic response, with a short initial rise in ventilation followed by a roll-off (Eden and Hanson, 1987; Bureau et al., 1984; McCooke and Hanson 1985; Teppema and Dahan, 2010). Shortly after birth, the initial rise in ventilation is less than the roll-off (Bonora et al., 1984; Elnazir et al., 1996; Eden and Hanson, 1987; Liu et al., 2006; Teppema and Dahan, 2010). Once adulthood is reached, the initial rise in ventilation in response to hypoxia is more pronounced and the roll-off becomes meniscal, resulting in a persistent rise in ventilation (Easton et al., 1986; Vizek et al., 1987; Maxova and Vizek, 2001; Pokorski and Antosiewicz, 2010; Teppema and Dahan, 2010). Previous studies have shown that adult mice lacking carotid bodies due to developmental atrophy do not hyperventilate and do not survive more than a couple days in hypoxic environments (Macías et al., 2014, 2018).

It is widely accepted in the literature that the hypoxic ventilatory response is sensed via the CB. This O_2 sensing ability of the CB depends on the presence of several subtypes of O_2-regulated K^+ channels present in glomus cell membranes whose open probability decreases during hypoxia exposure, thus triggering cell depolarization, Ca^{2+} influx via low- and high-threshold Ca^{2+} channels, and Ca^{2+}-dependent exocytotic transmitter release (López-Barneo et al., 2016). This well-recognized hypoxic signaling pathway in the CB is known as the membrane model of hypoxic chemotransduction. The sensitivity of CB glomus cells to hypoxia, as demonstrated by changes in cytosolic Ca^{2+} or exocytotic transmitter release as a function of environmental pO_2, follows a hyperbolic function parallel to the curve relating ventilation and arterial O_2 tension in animals (López-Barneo et al., 2016). Nevertheless, the contribution of central chemoreceptive respiratory neurons and glia to the hypoxic ventilatory response is controversial (Gourine and Funk, 2017).

HYPERCAPNIC VENTILATORY RESPONSE

Elevations in arterial pCO_2 are sensed by both the CB and central chemoreceptors to trigger an increase in ventilation. The ventilatory response to HC gas challenge has been studied considerably in rats and follows a definitive pattern from birth to adulthood, mainly the (1) ventilatory response to HC declines (i.e., shows roll-off) in the first postnatal week (Putnam et al., 2005; Stunden et al., 2001), (2) magnitudes of HC responses are at their lowest points during the second week of life, and (3) responses to HC increase substantially during the third week (i.e., as seen in P21 rats) toward those seen in adult rats (Putnam et al., 2005; Stunden et al., 2001).

Consistent with hypoxia, hypercapnia and secondarily, acidemia, depolarize the CB glomus cell membrane and trigger Ca^{2+}-dependent neurotransmitter release (Buckler and Vaughan-Jones, 1994; Zhang and Nurse, 2004). Glomus cells, along with central chemoreceptors, are activated by increases in arterial pCO_2 mainly as the result of intracellular acidification, an activity catalyzed by carbonic anhydrase (Nurse, 1990). Evidence has shown that pharmacological carbonic anhydrase inhibition diminishes hypercapnic cellular responses (Zhang and Nurse, 2004; Iturriaga et al., 1993) and systemic ventilatory activation (Wagenaar et al., 1998). Studies have also shown that CB glomus cell CO_2/H^+ sensing is the outcome of the redundant activation of various ion channel subtypes (e.g. voltage-dependent K^+ channels, maxi-K^+ channels, background TASK-like K^+ channels, L-type Ca^{2+} channels, acid-activated cationic channels, and an inwardly rectifying chloride channel (Zhang and Nurse, 2004, Summers et al., 2002; Buckler et al., 2000; Peers and Green, 1991; López-López et al., 1989; Tan et al., 2007; Petheo et al., 2001), therefore suggesting that the elimination of genes encoding certain channel types, such as TASK-like channels, may not have significant functional cellular or organismal effect when exposed to hypercapnia/acidosis since other channels are present to sense CO_2/H^+. For example, mice lacking TASK1 and TASK3 channels, which have been thought to play a major role in CO_2 and acid sensing, have a normal ventilatory response and glomus cell responsiveness to hypercapnia (Ortega-Sáenz et al., 2010).

HYPOXIC-HYPERCAPNIC VENTILATORY RESPONSE

Although individual activation of peripheral chemoreceptors (e.g., carotid bodies) by HX and HC, or central chemoreceptors by HC causes breathing to increase, the ventilatory response when HX and HC are given together (i.e., a hypoxic-hypercapnic (HH) gas exposure), arguably one that induces more physiologically relevant changes in arterial blood-gas chemistry, has received conflicting results. Three currently recognized hypotheses are that the interaction between simultaneous inputs from both central and peripheral chemoreceptors can be (1) positive in that one stimulus (e.g., HX or HC) augments the effects of the other in a synergistic/multiplicative manner; (2) negative in that the inputs are hypo-additive; or (3) additive in that when HX and HC are given together, one stimulus, HX or HC, simply adds to the other (Eldridge et al., 1981). Experiments have shown positive or synergistic interactions between HX and HC at the level of the CB in cats (Eyzaguirre and Lewin, 1961). In 2015, Wakai et al. exposed adult rats to HH gas challenge showing enhanced increases in respiration compared to HX exposures alone, suggesting that HX and HC ventilatory signaling may be synergistic in rats (Wakai et al., 2015). Additional studies in adult dogs shows that peripheral CB and central chemoreceptors are functionally inter-dependent, thus suggesting that activity and/or sensitivity of medullary chemoreceptors is determined by signaling from CB chemoreceptors and *vice versa* (Fencl et al., 1966; Smith et al., 1984, 2006, 2010, 2015; Dempsey, 2005; Blain et al., 2010; Forster and Smith, 2010). Conversely, other studies have data in cats and rabbits that shows the ventilatory responses to a constant CSN stimulation (mimicking HX exposure) decrease progressively as arterial pCO_2 rises (mimicking a HC exposure), suggesting a hypo-additive interaction between HX and HC (Eldridge, 1974; Kiwull et al., 1976). Moreover, an early study done in cats has shown that when HH is given, the interaction between central and peripheral chemoreceptors is additive in that one stimulus, HX or HC, simply adds to the other (Eldridge et al., 1981). Nonetheless, until this book project no whole animal studies have looked at the effects of HH gas challenge on ventilatory expression in juvenile (postnatal day 25) rats and in juvenile rats following bilateral transection of the carotid sinus nerve (CSNX).

LOSS OF CHEMOSENSORY AFFERENT INPUT INHIBITS COMPENSATORY VENTILATORY ADJUSTMENTS TO HYPOXIA AND HYPERCAPNIA

The loss of CSN innervation to the CB bilaterally in humans and most mammals results in blunted hypoxic and hypercapnic ventilatory responses at various developmental stages (Teppema and Dahan, 2010; Timmers et al., 2003; Limberg, 2018; Baby et al., 2018; Olson et al., 1988; Sapru and Krieger, 1977; Mouradian Jr. et al., 2012). In newborns, the initial peak in ventilation following HX challenge is significantly attenuated after bilateral CSNX (Bureau et al., 1985; Fung et al., 1996; Lowry et al., 1999a; Lowry et al., 1999b; Suguihara et al., 1994; Teppema and Dahan, 2010). In adult mammals, bilateral CSNX stimulates hypoventilation under room-air (Sapru and Krieger, 1977), and practically abolishes the ventilatory responses to HX and HC challenges (Baby et al., 2018; Olson et al., 1988; Sapru and Krieger, 1977; Mouradian Jr. et al., 2012; Cardenas and Zapata, 1983; Chalmers et al., 1967). Nonetheless, within a few weeks after bilateral CSNX, there is an obvious restoration of HX ventilatory responses (Bisgard et al., 1976; Smith and Mills, 1980; Martin-Body et al., 1985, 1986) that are correlated with a definite functional reorganization of central chemoreceptor reflex pathways, including alterations in ventilatory pattern and brainstem catecholaminergic activity (Roux et al., 2000). Other reported mechanisms involved in the recovery of HX ventilatory responses include (1) increased aortic body activity (Bisgard et al., 1980; Serra et al., 2002), (2) increased chemosensory input from secondary O_2 sensing tissue in the head/neck (Serra et al., 2002, Martin-Body, et al., 1986; Lowry et al., 1999a), and (3) regeneration of sensory terminals of the CSN, and/or plastic changes in the central respiratory control network including increased activation of O_2 sensing neurons (Smith and Mills, 1980; Lowry et al., 1999b; Serra et al., 2001; Mitchell and Johnson, 2003; Forster, 2003; Teppema and Dahan, 2010).

SYMPATHETIC NERVOUS SYSTEM

So far, we have talked about the structure and function of the carotid body, and carotid sinus nerve, and how some diseases (e.g., hypertension, chronic heart failure or sleep apnea) are linked to dysfunctions in the CB and hyperexcitability of the CSN. We also defined the widespread location and function of central chemoreceptors and how central chemoreception is interdependent with peripheral chemoreceptor activation. In addition, we described the known mechanisms behind the ventilatory responses to hypoxia, hypercapnia and hypoxia-hypercapnia, and how loss of chemosensory afferents blunts these ventilatory responses. In this final section we will focus on the sympathetic nervous system and its important role in modulating the activity of certain respiratory-related structures, such as the CB, in response to hypoxia.

The sympathetic nervous system is a division of the autonomic nervous system and supplies efferent axons to all viscera of the body in order to constantly adjust organ function during periods of stress, anxiety, physical activity, fear, or excitement. The sympathetic nervous system arises in the central nervous system (i.e., spinal cord) and innervates target tissue by a two-synapse pathway (Boron and Boulpaep, 2012). The cell bodies of the sympathetic preganglionic neurons are located in the intermediolateral cell column between the dorsal and ventral horns of the thoracic and upper lumbar spinal cord between levels T1 and L3 (Boron and Boulpaep, 2012). These sympathetic preganglionic neurons send projections out of the spinal cord to make synapses with sympathetic postganglionic neurons located in peripheral ganglia (Boron and Boulpaep,

2012). Axons from these sympathetic postganglionic neurons then project to their specific target tissue.

SUPERIOR CERVICAL GANGLIA PROVIDES SYMPATHETIC POSTGANGLIONIC INPUT TO THE CAROTID BODY, BRAINSTEM, TONGUE, AND UPPER AIRWAY

The bilateral superior cervical ganglia (SCG) are the most rostral sympathetic ganglia, arising from the fusion of cervical ganglia (C)1 to C4, and supplying the head and neck. The cell bodies of postganglionic sympathetic neurons (Llewellyn-Smith et al., 1998; Tang et al., 1995a,b,c; Rando et al., 1981) and small intensely fluorescent (SIF) cells present (Takaki et al., 2015; Zaidi and Matthews, 2013; McDonald, 1983a,b) within the SCG receive their preganglionic input from thoracic (T) spinal cord (T1-T4) nerves that course within the cervical sympathetic chain (CSC). The sympathetic preganglionic neurons traveling in the CSC release acetylcholine and stimulate nicotinic (N_2) receptors on cells in the SCG (Boron and Boulpaep, 2012). The activated principal SCG postganglionic neurons then release norepinephrine onto various adrenergic receptor subtypes present on target tissues (Boron and Boulpaep, 2012).

Two main trunks exit the SCG and are known as the external (ECN) and internal (ICN) carotid nerves (Savastano et al., 2010; Asomoto, 2004; Buller and Bolter, 1997; Bowers and Zigmond, 1979). An additional branch off the ECN, called the ganglioglomerular nerve (GGN) sends principal SCG postganglionic fibers to innervate: (1) numerous structures in the CB, including the vasculature, chemoafferent receptor terminals, and type 1 glomus cells (Savastano et al., 2010; Asamoto, 2004; Ichikawa, 2002 Torrealba and Claps, 1988; Verna et al., 1984; McDonald, 1983a,b; McDonald and

Mitchell, 1981; Bowers and Zigmond, 1979; Zapata et al., 1969; Biscoe and Purves, 1967), and (2) terminals of baroreceptor afferent nerves within the carotid sinus (Buller and Bolter, 1993; Felder et al., 1983; Bolter and Ledsome, 1976; Rees, 1967; Floyd and Neil, 1952). In addition to SCG postganglionic innervation of the CB and carotid sinus, the SCG also sends projections to nuclei within the brainstem, such the NTS and hypothalamus (Mathew, 2007; Hughes-Davis et al., 2005; Esquifino et al., 2004; Westerhaus and Loewy, 1999; Wiberg and Widenfalk, 1993; Saavedra, 1985; Gallardo et al., 1984; Cardinali et al., 1982, 1981a,b) as well as the tongue and upper airway (Oh et al., 2006; Wang and Chiou, 2004; Hisa et al., 1999; O'Halloran et al., 1996, 1998; Kummer et al., 1992; Flett and Bell, 1991).

For the rest of this section, we will focus on the role of sympathetic innervation of the main peripheral chemoreceptor, the CB. SCG postganglionic innervation is known for playing an influential part in affecting CB chemosensitivity largely (EYZAGUIRRE, and LEWIN, 1961; Purves, 1970), but not solely (O'Regan, 1975; Verna et al., 1984), by modulating CB blood flow (Kumar and Prabhakar, 2012). For instance, activation of SCG postganglionic nerve fibers indirectly activates CB glomus cells by constricting arterioles via activation of α_1-adrenoceptors and dopamine receptors present on smooth muscle cells, therefore mimicking a hypoxic environment for glomus cells (Llados and Zapata, 1978; Majcherczyk et al., 1980; Matsumoto et al., 1981; Yokoyama et al., 2015). In addition, co-release of neuropeptide Y from SCG postganglionic nerve terminals also reduces blood flow within the CB (Potter and McCloskey, 1987). Conversely, stimulation of SCG postganglionic neurons can directly actives CB glomus cells by activation of β-

adrenoceptors and dopamine receptors present on these cells (Eldridge and Gill-Kumar, 1980; Lahiri et al., 1981; Gonsalves et al., 1984). Moreover, the effect of stimulating SCG postganglionic fibers on CB afferent discharge is most often excitatory although it can be inhibitory (Matsumoto et al., 1986; Mitchell and McCloskey, 1974; O'Regan, 1981). For example, endogenous norepinephrine indirectly inhibits carotid body chemoafferent activity (Overholt and Prabhakar, 1999) via inhibition of excitatory neurotransmitter release from glomus cells by decreasing the magnitude and rate of activation of the macroscopic Ca^{2+} currents (Overholt and Prabhakar 1999). Interestingly, a study has also shown that if the carotid body is stimulated *in vivo*, the GGN exhibits inhibitory activity, mediated by nitric oxide (NO) generated from neuronal NO synthase (nNOS) (Campanucci, and Nurse, 2007), that leads to a reduction in CSN chemoafferent discharge (Majcherczyk et al., 1974).

STIMULATION OF CSC PREGANGLIONIC FIBERS AND PRINCIPAL SCG POSTGANGLIONIC NEURONS, AS WELL AS SIF CELLS, MODULATE CAROTID BODY ACTIVITY DURING NORMOXIA AND HYPOXIA

Sympathetic innervation in the CB influences resting (i.e., baseline normoxic) activity of glomus cells and chemosensory afferent fibers within the CSN (Prabhakar, 1994). Numerous studies have reported a variety of responses upon application of neurotransmitters naturally released by CSC and GGN nerve terminals (e.g. dopamine, norepinephrine, and neuropeptide Y) to both *in vivo* and *in vitro* CB preparations during normoxia, including (1) a biphasic pattern consisting of initial brief bursts in CSN activity and then long-lasting inhibition (Bisgard et al., 1979), (2) a biphasic pattern consisting of initial brief decreases in CSN activity and then long-lasting excitation (Matsumoto et al., 1981), (3) indirect excitation of CB glomus cells via constriction of arteriolar blood flow in the CB (Potter et al., 1987; Yokoyama et al., 2015), (4) direct activation of glomus cells and/or CSN chemosensory afferents (Matsumoto et al., 1980; Lahiri et al., 1981; Milsom et al., 1983; Heinert et al., 1995; Pang et al., 1999), and (5) direct inhibitory action of glomus cells and/or CSN chemosensory afferents (Zapata et al., 1969; Zapata, 1975; Llados and Zapata, 1978. Mills, et al., 1978; Folgering et al., 1982; Kou et al., 1991; Pizarro et al., 1992; Bisgard et al., 1993; Prabhakar et al., 1993; Ryan et al., 1995; Almaraz, et al., 1997; Overholt and Prabhakar, 1999).

Moreover, there is substantial evidence showing that (1) sympathetic innervation to the CB potentiates and suppresses activity of CB glomus cells during hypoxia (Prabhakar, 1994), (2) principal SCG postganglionic neurons are not hypoxia sensitive (Bernardini et al., 2020; Gao et al., 2019; Buckler and Turner, 2013; Nunes et al., 2010, 2012; Rigual et al., 1999), and (3) subpopulations of SCG cells and especially SIF cells (i.e., dopaminergic interneurons, comprising only 3% of the total SCG cell population (Libet, 1970; Eccles and Libet, 1961)) are hypoxia sensitive (Nunes et al., 2010, 2012; Strosznajder et al., 1997; Dinger et al., 1993; Brokaw and Hansen, 1987; Hanson et al., 1986). Authors have shown that CSC and GGN activity increases during hypoxic challenge, suggesting the release of neurotransmitters, such as dopamine, norepinephrine, and neuropeptide Y onto CB glomus cells (Matsumoto et al., 1986, 1987; Lahiri et al., 1986, Yokoyama et al., 2015). Conversely, previous studies have shown that intra-carotid artery infusions of norepinephrine attenuate hypoxic excitation of the CB through activation of α_2-

adrenoceptors (Kou et al., 1991; Pizarro et al., 1992; Prabhakar et al., 1993; Almaraz et al., 1997). In addition, there is evidence showing that activation of the GGN decreases the hypoxic response of chemoreceptors within the cat carotid body (McQueen et el., 1989).

As mentioned, (see references above) SIF cells in the SCG possess O_2 sensing properties and directly respond to hypoxic challenges. Dinger et al (1993) demonstrated that SCG SIF cells are selectively activated by hypoxic stimuli, and that the neurochemical responses of these SIF cells are similar to those observed by chemosensory CB glomus cells. In addition, Dalmaz et al (1993) revealed that SIF cells show morphological and biochemical similarities to the well-known hypoxia sensitive CB glomus cells, and McDonald and Blewett (1981) showed that SIF cells reside in clusters near fenestrated capillaries, similar to the way CB glomus cells form clusters near fenestrated capillaries. Moreover, there is compelling evidence in strong favor of SIF cells being the sole hypoxia-sensitive cell type in the SCG despite conflicting results (see references above) that principal SCG cells may also contain hypoxia sensing properties. This evidence includes that (1) the morphological plasticity following prolonged hypoxia in SIF cells of the SCG is similar in CB glomus cells containing dopamine (Soulier et al. 1997; Dalmaz et al., 1993), (2) similar to CB glomus cells, SIF cells receive both afferent and efferent innervation (McDonald, 1983a,b; Zaidi and Matthews, 2013; Takaki et al. 2015), (3) a combination of immunohistochemistry and retrograde labeling has allowed identification of three subtypes of SIF cells in the SCG, and again similar to CB glomus cells, most afferent endings on SIF cells have cell bodies located in the petrosal ganglion (Takaki et al., 2015), and (4) afferent endings on SIF cells express purinergic P2X3 receptors (Takaki et al., 2015), thus revealing a sensory pathway analogous to carotid body chemoreceptors, in which ATP acting on P2X3-containing chemoafferent terminals of the CSN is a major mechanism for CB chemosensory transduction (Nurse, 2014).

DISEASES ASSOCIATED WITH DISRUPTED ACTIVITY OF THE CSC-SCG COMPLEX

Clinical subjects suffering from chronic T1-T4 spinal cord injury present with cardiorespiratory disturbances consistent with diminished activity of the CSC-SCG complex (DiMarco et al., 2009; Heutink et al., 2014; Sankari et al., 2014, 2019; Berlowitz et al., 2016; Hachmann et al., 2017; Shin et al., 2019) including, enhanced peripheral chemoreflex sensitivity, which is a major cause of sleep-disordered breathing in these patients (Tester et al., 2014; Bascom et al., 2016). Evidence in rats has shown that mid-thoracic spinal cord injury is associated with enhanced cardiac sympathetic activity and cardiac sympathetic hyperinnervation that increases the susceptibility to life-threatening arrhythmias (Rodenbaugh et al., 2003; Collins et al., 2006; Lujan and DiCarlo 2007; Lujan et al., 2009, 2010, 2012, 2014). Studies in rats have also confirmed that hypoactivity of the CSC greatly enhances the likelihood of stroke in hypertensive models (Sadoshima et al., 1981; Sadoshima et al., 1983; Werber and Heistad, 1984). However, until this book project, no *in vivo* studies have characterized the effects of transection of the CSC or surgical removal of the SCG on breathing or the ventilatory responses that occur during hypoxic gas exposure. Nonetheless, the effects of bilateral transection of the CSC (CSCX) or bilateral removal of the SCG (SCGX) have been investigated on a variety of other physiological functions/variables in the mouse (Krieger, et al., 1976; García et al., 1988; Kawaja and Crutcher, 1997; Pankevich et al., 2003a,b; Karlsen et al., 2013; Lindborg et al.,

2018; Ziegler et al., 2018). For instance, García et al (1988) showed a significant increase of glucose-stimulated insulin release in pancreas cells *in vitro* in mice 14-20 h after SCGX, and an inhibition of glucose-stimulated insulin release at the 96 h timepoint after SCGX, and Lindborg et al (2018) revealed that different subpopulations of neutrophils exist in the SCG after unilateral transection of the CSC.

CHAPTER II: THE ROLE OF CAROTID SINUS NERVE INPUT IN THE HYPOXIC-HYPERCAPNIC VENTILATORY RESPONSE IN JUVENILE RATS

INTRODUCTION

The detection of oxygen (O_2), carbon dioxide (CO_2) and proton (H^+) levels in arterial blood is crucial for the control of breathing (Prabhakar, 2000; Teppema and Dahan, 2010; Guyenet and Bayliss, 2015; Guyenet et al., 2019). Within seconds, arterial hypoxia (low pO_2), hypercapnia (high pCO_2) and/or acidosis (high H^+ levels) stimulate breathing, which is essential for maintaining proper oxygenation of tissues. This ventilatory adjustment depends critically on the O_2-sensing ability of the carotid bodies (CBs) and respiratory control brainstem nuclei (Prabhakar, 2000). The CBs are vascularized organs that detect changes in arterial blood pO_2, pCO_2 and pH. Decreases in pO_2 and pH, and/or elevations in pCO_2 stimulate chemosensitive glomus (type I) cells in the CBs to release neurotransmitters that excite closely apposed nerve terminals of chemoafferent fibers within the carotid sinus nerve (CSN), thereby causing an increase in firing of these afferents (Gonzalez et al., 1994; Eyzaguirre and Zapata, 1984; Kikuta et al., 2019; Peers and Buckler, 1995; Prabhakar, 2000; Tse et al., 2012). CSN chemoafferents then relay these signals to the commissural nucleus tractus solitarius (cNTS) within the brainstem (Song et al., 2011), and increased inputs from these chemoafferent fibers trigger downstream responses to restore arterial blood gas status and hemodynamic homeostasis (Nurse, 2005; Nurse, 2010; Prabhakar, 2000).

The ventilatory responses to hypoxic (HX) and hypercapnic (HC) gas challenges differ in new-born compared to adult animals. In neonates, the HX ventilatory response appears as a biphasic response comprising of a short initial rise in ventilation (augmentation phase) quickly followed by a secondary roll-off, generally to a level below baseline normoxia (depressant phase) (Eden and Hanson, 1987; Bureau et al., 1984; McCooke and Hanson 1985; Teppema and Dahan, 2010). Shortly after birth in species such as the pig, cat and rat, the initial rise in minute ventilation (MV) is smaller than the secondary roll-off, resulting in lower baseline ventilation values compared to those under normoxia (Bonora et al., 1984; Elnazir et al., 1996; Eden and Hanson, 1987; Liu et al., 2006; Teppema and Dahan, 2010). Once adulthood is reached, the initial rise in ventilation in response to hypoxia is more pronounced and the secondary depression becomes smaller, resulting in a sustained rise in MV (Easton et al., 1986; Vizek et al., 1987; Maxova and Vizek, 2001; Pokorski and Antosiewicz, 2010; Teppema and Dahan, 2010). The ventilatory response to HC gas challenge during development has been studied extensively in rats and follows a definitive pattern, namely (1) the ventilatory response to HC declines (i.e., shows roll-off) during the challenge in the first postnatal week (Putnam et al., 2005; Stunden et al., 2001), (2) the magnitudes of HC responses are at their lowest points during the second week, and (3) the HC response increases during the third week (i.e., as seen in P21 rats) toward those seen in mature (adult) rats (Putnam et al., 2005; Stunden et al., 2001). The ventilatory response to combined HX and HC (HH) gas challenges (arguably those that induce more physiologically relevant changes in arterial blood-gas chemistry) has not been studied in neonatal rats. However, adult rats exposed to HH gas challenge show enhanced increases in ventilation compared to HX exposures alone, although whether the responses were greater than HC gas challenge alone was not determined (Wakai et al., 2015). Nonetheless, the data strongly suggest that HX and HC ventilatory signaling may be synergistic in rats (Wakai et al., 2015).

The loss of CSN innervation to the CB greatly inhibits compensatory respiratory adjustments in response to HX and HC challenges in neonatal and adult animals. In newborns, the initial rise in ventilation (augmentation phase) in response to acute HX challenge is significantly attenuated after bilateral CSNX (Bureau et al., 1985; Fung et al., 1996; Lowry et al., 1999a; Lowry et al., 1999b; Suguihara et al., 1994; Teppema and Dahan, 2010). In adult mammals, bilateral CSNX induces hypoventilation under room-air

(Sapru and Krieger, 1977), and diminishes the ventilatory responses to HX and HC challenges (Baby et al., 2018; Olson et al., 1988; Sapru and Krieger, 1977; Mouradian Jr. et al., 2012; Cardenas and Zapata, 1983; Chalmers et al., 1967). Within a few weeks after bilateral CSNX there is a noticeable recovery of HX ventilatory responses (Bisgard et al., 1976; Smith and Mills, 1980; Martin-Body et al., 1985, 1986) that are associated with a distinct functional reorganization of central chemoreceptor reflex pathways, including changes in ventilatory pattern and brainstem catecholaminergic activity (Roux et al., 2000). Other reported mechanisms involved in restoration of HX ventilatory response include (1) enhanced aortic body activity (Bisgard et al., 1980; Serra et al., 2002), (2) enhanced chemosensory input from secondary glomus tissue in the head/neck (Serra et al., 2002, Martin-Body, et al., 1986; Lowry et al., 1999a), and (3) regeneration of sensory terminals of the carotid sinus nerve, and/or plastic changes in the central respiratory neuronal network involving activation of oxygen-sensing neurons (Smith and Mills, 1980; Lowry et al., 1999b; Serra et al., 2001; Mitchell and Johnson, 2003; Forster, 2003; Teppema and Dahan, 2010).

Along with CB chemoreceptors, mammalian species also possess central chemoreceptors, such as the retrotrapezoid nucleus (RTN), which play a definitive role in breathing modulation, and are key responders to elevated blood pCO_2 (Nattie, 2006; Guyenet and Bayliss, 2015; Guyenet et al., 2019). The CBs detect arterial pCO_2 (and pH) and monitor alveolar ventilation, whereas central chemoreceptors detect interstitial fluid pH and monitor the balance of arterial pCO_2, cerebral blood flow, and cerebral metabolism (Fencl et al., 1966; Dempsey, 2005; Xie et al., 2005; Smith et al., 2006). The time course of responses varies between central and peripheral chemoreceptors, with CBs providing the rapid response to changes in arterial pCO_2, while central chemoreceptors provide most of the steady-state response (Smith et al., 2006; Gaston et al., 2014). Studies in conscious humans with intact or surgically resected CBs revealed that the fast time constant in response to brief CO_2 pulses was absent in patients with CB resection, therefore providing strong support for the rapid response function of the CBs (Fatemian et al., 2003).

Although activation of peripheral or central chemoreceptors causes breathing to increase, the type of interaction between the two systems remains unresolved. It has been hypothesized that when a HX and HC stimulus are given together, the interaction between the input from both central and peripheral chemoreceptors can be (1) positive in that one stimulus (e.g. HX or HC) augments the effects of the other in a synergistic/multiplicative, manner or (2) negative in that the inputs are hypo-additive (Eldridge et al., 1981). Studies have shown positive or synergistic interaction between HX and HC at the level of the CB (Eyzaguirre and Lewin, 1961). Other studies have shown that ventilatory responses to a constant CSN stimulation (mimicking HX exposure), decreased progressively as arterial pCO_2 rose (mimicking a HC exposure), thus suggesting a hypo-additive interaction between the two stimuli (Eldridge, 1974; Kiwull et al., 1976).

Despite evidence that (1) neonatal and adult rats display different HX and HC ventilatory responses, and (2) bilateral CSNX blunts HX and HC ventilatory responses in adult and newborn rats, the effects of bilateral CSNX on HX, HC or HH gas challenge in juvenile SD rats are unknown. Moreover, the vital question of how HX and HC signaling pathways interact with one another and the role of the CBs in this interaction is not well characterized at any age. To address this, we examined the ventilatory responses elicited by a single 5-min episode of HX, HC or HH gas challenge in conscious, freely-moving Sprague Dawley juvenile (postnatal age 25, P25) rats that underwent prior (4 days earlier) sham-operation (SHAM) or bilateral CSNX to disrupt the chemoreceptor reflex pathway. Each gas challenge was given to separate groups of rats (e.g., HX challenge had a group of SHAM and CSNX rats that was separate from the HC and HH challenge) in order to prevent any confounding influences that one gas challenge may have on the other. Our novel findings demonstrating that CSNX blunts the HX-, HC- and HH-induced ventilatory responses provide the first *in vivo* data showing that hyperventilation in response to these challenges in juvenile rats is dependent on CB chemoafferent signaling. Our results also show that the ventilatory responses to HX alone and HC alone were simply additive in SHAM rats and that this additivity was lost in CSNX rats, suggesting that peripheral CB chemoafferent activity plays an essential role in determining the interactions between peripheral and central chemoafferent pathways controlling ventilation.

RESULTS

RESTING PARAMETERS

A summary of the descriptors and baseline ventilatory parameters for SHAM and CSNX rats is provided in **Table 1**. The ages and body weights of the SHAM and CSNX groups for the HX, HC, and HH gas challenges were statistically similar to one another (upper rows of **Table 1**). As such, no corrections for differences in body weights were applied to the ventilatory data. The ventilatory parameters shown in the middle rows of **Table 1** are those recorded throughout the 15-min baseline period, and show that fR, VT and MV, in each CSNX group for the HX (poikilocapnic), HC and HH gas challenge were significantly smaller compared to those in the respective SHAM group (P < 0.05, CSNX *versus* SHAM in HX, HC and HH for all comparisons). The ventilatory values in the bottom rows of **Table 1** represent those recorded immediately (2-3 min) before each HX, HC and HH gas challenge in SHAM and CSNX groups (i.e., Pre-values). Pre-values were significantly depressed for fR, VT and MV in each CSNX group (i.e., each CSNX group for HX, HC and HH gas challenge) compared to the respective SHAM group.

EXAMPLE TRACES RECORDED AT BASELINE AND DURING HX, HC, OR HH GAS CHALLENGES

Representative sections of plethysmography traces from SHAM and CSNX rats are shown in **Figure 3**. In response to HX, the fR in the SHAM rat increased from 102 to 144 breaths/min (+42 breaths/min, +41%). The increase in fR in the CSNX rat was smaller (from 102 to 108 breaths/min, +6 breaths/min). In response to HC, fR in the SHAM rat increased from 96 to 144 breaths/min (+48 breaths/min). The increase in fR in the CSNX rat was similar (from 108 to 162 breaths/min, +54 breaths/min, +50%). In response to HH, fR in the SHAM rat increased from 102 to 186 breaths/min (+84 breaths/min). The increase in fR in the CSNX rat was smaller (from 102 to 138 breaths/min, +36 breaths/min, +35%). The increase in fR to HH in the SHAM rat (+84 breaths/min) was similar to simple addition of the HX and HC responses (42 + 48 = 90 breaths/min), whereas the HH response in the CSNX rat (+36 breaths/min) was much smaller than the simple addition of the HX and HC responses (6 + 54 = 60 breaths/min).

COMPARISON OF THE HX-, HC- AND HH-INDUCED RESPONSES TO THE MAXIMAL ATTAINABLE VALUES

The detailed changes in ventilatory responses to HX, HC and HH gas challenges will be described below. It is important to note that these challenges, including HH, did not produce changes in ventilatory parameters that approached the maximal attainable responses recorded in these rats throughout the entire study. Specifically, the Pre-values, the maximal values recorded during the entire study (maximum), and the peak values obtained during the HX, HC or HH are shown in **Table 2**. For fR (top rows), VT (middle rows) and MV (bottom rows), the delta values for the gas challenge responses were all smaller than the delta values for the maximal attainable responses in the CSNX group compared to the SHAM (P < 0.05, for all comparisons).

FREQUENCY OF BREATHING (f_R) IN SHAM AND CSNX RATS DURING HX, HC OR HH GAS CHALLENGES

The top panels of **Figure 4** summarize fR values of SHAM and CSNX rats before, during and after exposure to a 5 min HX, HC or HH gas challenge. In SHAM rats (upper left panel), HX and HC elicited robust increases in fR that appeared to display roll-off toward the end of the challenge. HH in SHAM rats elicited a robust increase in fR that did not display roll-off. In CSNX rats (upper right panel), HX elicited an increase in fR that was subject to a more pronounced roll-off than in SHAM rats. The increase in fR elicited by HC in CSNX rats looked comparable to that in SHAM rats, whereas the increase in fR elicited by HH challenge looked substantially smaller in CSNX rats than SHAM rats. The bottom panels of **Figure 4** summarize the arithmetic changes in fR that occurred throughout the actual HH protocol, and simple addition of HX and HC values (HX+HC, SEM values presented as 10% of the mean to provide perspective). The actual HH values in SHAM rats (bottom left panel) were smaller than the HX+HC values during the initial phases of the HH gas challenge compared to HX+HC, and the initial seconds upon return to room-air. In SHAM rats, the maximal attainable changes in fR during the entire HH protocol was +224 ± 19 breaths/min (**Table 2**), which was much higher than the changes in the HH gas challenge (+126 ± 10 breaths/min) (**Table 2**). In SHAM rats, the summed changes in HX and HC responses (+80 ± 9 and +105 ± 18 breaths/min, respectively; +185 breaths/min upon addition) was smaller than the maximal attainable changes in fR during the entire HH protocol (+224 ± 19 breaths/min). In contrast to SHAM rats, the HH and HX+HC responses in CSNX rats (bottom right panel) were similar to one another. For CSNX rats, the maximal attainable changes in fR during the entire HH protocol (+271 ± 14 breaths/min) (**Table 2**), was again substantially higher than those observed during the HH gas challenge (+105 ± 15 breaths/min) or those obtained by summation of the HX and HC responses (+64 ± 9 and +84 ± 14 breaths/min, respectively; +148 breaths/min upon addition).

The top panels of **Figure 5** compare the total increases in fR that occurred during the single HX, HC or HH gas challenge in SHAM and CSNX rats. All three challenges elicited substantial increases in fR from Pre-values. The increase in fR elicited by HX was smaller in CSNX rats than in SHAM rats. The increase in fR during HC were not statistically smaller in CSNX rats compared to SHAM rats. The increase in fR during HH in CSNX rats was smaller than in SHAM rats. **Table 3** shows the arithmetic differences between the total 5 min HX, HC and HH fR responses in SHAM and CSNX rats, with the (-) negative values meaning that the responses in CSNX rats were smaller than in SHAM rats. The differences in responses of CSNX and SHAM rats for the additive HX+HC values are also shown. The actual fR differences in HH responses between SHAM and CSNX rats were far smaller than predicted by the addition of HX and HC (HX+HC) responses. The bottom panels of **Figure 5** summarize the total fR responses elicited by HX, HC and HH in SHAM rats (bottom left panel) and CSNX rats (bottom right panel). The HX+HC values are also shown, with the values above the columns noting the percentage difference between the actual HH values and HX+HC values. In SHAM rats, the fR response elicited by the actual HH gas challenge was smaller than the summed HX+HC values. In the CSNX rats, the fR response elicited by the actual HH gas challenge was similar to the summed HX+HC values, suggesting that fR is less than additive in SHAM rats and mostly additive in CSNX rats.

Rapid responses to HX, HC and HH would be expected to involve CB chemoafferents. As such, we focused on the fR responses that occurred during the first 60 sec of the HX, HC and HH gas challenge in SHAM and CSNX rats. The changes in fR that occurred during the first 60 sec of the HX, HC and HH gas challenge are shown in the top panels of **Figure 6**. For SHAM rats (top left panel), the initial increases in fR elicited by HX were less than those elicited by the HC or HH, whereas there were minimal differences between HX, HC and HH responses in CSNX rats (top right panel). Overall, the responses during the first 60 sec of HX, HC and HH were similar in SHAM and CSNX rats. Next, we wanted to determine whether the changes in magnitude of the actual fR response during the first 60 sec of the HH gas challenge was simply additive of those elicited by the HX and HC alone. As seen in the middle panels of **Figure 6**, the actual HH values in SHAM rats (middle left panel) were substantially less than those of the summed HX+HC values. The actual HH values in CSNX rats (middle right panel) were somewhat less than those of the added HX+HC values. The bottom panels of **Figure 6** show the total fR responses that occurred during the first 60 sec of HX, HC and HH gas challenge in SHAM and CSNX rats. Again, for SHAM rats (bottom left panel), the increase in fR during the first 60 sec elicited by HH was less than would be expected by simple addition of the HX and HC responses. This was also true for CSNX rats (bottom right panel). The clear impression given by the data is that there is a negative interaction between HX and HC signaling pathways during the first 60 sec of a HH gas challenge, and this interaction is independent of the presence or absence of the CB-CSN complex.

TIDAL VOLUME (VT) IN SHAM AND CSNX RATS DURING HX, HC OR HH GAS CHALLENGES

The top panels of **Figure 7** summarize the VT values recorded in SHAM and CSNX rats before, during and after exposure to 5 min HX, HC or HH gas challenges. In SHAM rats (top left panel), HX and HC gas challenge elicited robust and sustained increases in VT that were similar to one another. HH gas challenge also elicited a sustained increase in TV that appeared greater than those elicited by HX or HC. In contrast, the increases in VT elicited by HX, HC and HH gas challenges in CSNX rats (top right panel) were similar to one another, and all smaller than those in SHAM rats. The bottom panels of **Figure 7** summarize the arithmetic changes in VT that occurred throughout the actual HH and the simple addition of HX and HC values (HX+HC). In SHAM rats (bottom left panel), the actual

HH values and additive HX+HC values were remarkably similar to one another. The maximal attainable changes in VT in the SHAM rat during the HH protocol, was +0.74 ± 0.04 ml (**Table 2**), which was higher than the changes during the HH gas challenge (+0.62 ± 0.03 ml), but equal to the sum of the HX and HC (HX+HC) responses (+0.37 ± 0.03 ml and +0.37 ± 0.03 ml, respectively; +0.74 ml upon addition). In contrast, CSNX rats (bottom right panel), the actual HH VT values were substantially smaller than the additive HX+HC values. The maximal attainable changes in VT in CSNX rats during the HH protocol, was +0.45 ± 0.03 ml (**Table 2**), which was higher than the changes during the HH gas challenge (+0.32 ± 0.02 ml), and lower than the sum of HX and HC (HX+HC) responses (+0.27 ± 0.03 ml and +0.30 ± 0.02 ml, respectively; +0.57 ml upon addition).

The top panel of **Figure 8** compares the total increases in VT that occurred during the single HX, HC or HH gas challenge in SHAM and CSNX rats. The total increases in VT elicited by HX were significantly smaller in CSNX rats than SHAM rats, whereas the VT responses during HC were similar in both groups. The total increases in VT elicited by HH were markedly smaller in CSNX rats than in SHAM rats. As shown in **Table 3**, the arithmetic differences in VT between SHAM and CSNX rats for HX of -2.01 ml and for HC of -0.43 ml would add to -2.44 ml, whereas the difference for the actual HH was -4.65 ml, suggesting that the actual VT difference in HH responses between SHAM and CSNX rats were far greater than predicted by addition of HX and HC (HX+HC) responses. The bottom panels of **Figure 8** summarize the total VT responses elicited HX, HC and HH in SHAM rats (bottom left panel) and CSNX rats (bottom right panel). The HX+HC values are also shown with figures above the HX+HC columns showing the percentage difference between the actual

HH values and the HX+HC values. In SHAM rats, the VT responses elicited by HX and HC appeared to be additive since the actual HH values were closely aligned with the computed HX+HC values. In contrast, the VT responses between HX and HC changed markedly in CSNX rats, such that the HX, HC and HH responses were similar to one another, and the actual HH responses were markedly lower than the added HX+HC responses. Therefore, the additive effects of HX and HC signaling pathways were markedly diminished in rats with bilateral CSN transection.

As mentioned, the rapid responses to HX, HC and HH would be expected to involve CB chemoafferents, and therefore we compared the changes in VT that occurred within the first 60 sec of the HX, HC and HH gas challenges in SHAM and CSNX rats. The changes in TV that occurred during the first 60 sec of the HX, HC and HH gas challenges are shown in the top panels of **Figure 9**. For SHAM rats (top left panel), the increases in VT elicited by HC were smaller than those elicited by HX or HH. The magnitude of the initial VT responses elicited by HX, HC and HH in CSNX rats (top right panel) were overall smaller than those in SHAM rats. Unlike the SHAM rats, HX, HC and HH elicited similar initial VT responses in CSNX rats. As seen in the middle panels of **Figure 9**, the magnitudes of the actual VT responses during the first 60 sec of the HH gas challenge were, in general, simply additive of those elicited by HX and HC alone in both SHAM and CSNX rats, except at the 60 sec time-point. The bottom panels of **Figure 9** summarize the total VT responses that occurred during HX, HC and HH in SHAM (bottom left panel) and CSNX (bottom right panel) rats. Again, for SHAM rats, the initial increases in VT elicited by HH were about that expected by simple addition of HX and HC responses. In contrast, the initial VT responses elicited by HH in CSNX rats were markedly less than expected from simple addition of HX and HC responses. Overall, and in contrast to the fR responses, the VT 60 sec data suggests that there is a positive interaction between HX and HC signaling pathways during the first 60 sec of a HH gas challenge, and that this interaction is dependent on the presence of the CB-CSN complex.

MINUTE VENTILATION (MV) IN SHAM AND CSNX RATS DURING HX, HC OR HH GAS CHALLENGES

The top panels of **Figure 10** summarize the MV values recorded in SHAM and CSNX rats before, during and after exposure to a 5 min HX, HC or HH gas challenge. In SHAM rats (top left panel), the HX and HC challenges elicited robust increases in MV that were relatively similar to one another, except there was a more robust roll-off during HX than HC. The increases in MV during HH were markedly greater than those during HX or HC. HX, HC and HH also elicited reproducible increases in MV in CSNX rats (top right panel) that appeared substantially smaller than those in SHAM rats. In CSNX rats, the increases in MV elicited by HH were similar to those elicited by HC, but greater than those elicited by HX. The bottom panels of **Figure 10** summarize the arithmetic changes in MV that occurred during the actual HH gas challenge and from simple addition of HX and HC gas challenge values (HX+HC). The actual HH values and additive HX+HC values were remarkably similar in SHAM rats (bottom left panel) and CSNX rats (bottom right panel). The maximal attainable changes in MV in SHAM rats during the HH protocol was +190 ± 6 ml/min (**Table 2**), which was much higher than the changes during the HH gas challenge (+168 ± 4 ml/min) (**Table 2**), but similar to the sum of the HX and HC (HX+HC) gas challenge responses (+84 $\pm$ 5 ml/min and +99 $\pm$ 13 ml/min, respectively; +183 ml/min upon addition). The maximal attainable changes in MV in CSNX rats during the HH protocol was +138 $\pm$ 6 ml/min (**Table 2**), which was substantially higher than those observed during the HH gas challenge (+85 $\pm$ 6 ml/min), and somewhat higher than the sum of the HX and HC (HX+HC) gas challenge responses (+46 $\pm$ 5 ml/min and +70 $\pm$ 5 ml/min, respectively; +116 ml/min upon addition).

The top panels of **Figure 11** compare the total increases in MV that occurred during the single HX, HC or HH gas challenge in SHAM and CSNX rats. The total increases in MV elicited by HX and HC were significantly smaller in CSNX compared to SHAM rats. The total increases in MV elicited by HH were also significantly smaller in CSNX than SHAM rats. **Table 3** shows that the arithmetic differences in MV between SHAM rats and CSNX rats for HX of -667 ml/min and for HC of -635 ml/min, adds to -1302 ml/min, and the difference in MV for the actual HH gas challenge was -1448 ml/min, suggesting that the MV difference in HH responses between SHAM and CSNX rats was similar to that predicted by addition of the HX and HC (HX+HC) responses. The bottom panels of **Figure 11** summarize the total MV responses elicited by the HX, HC and HH gas challenges in the SHAM rats (bottom left panel) and CSNX rats (bottom right panel). The summed HX+HC values are displayed with figures above the HX+HC columns showing the percentage difference between the actual HH values and HX+HC values. It seems evident that the MV response during HH was simply equal to the sum of the HX and HC (HX+HC) responses in SHAM rats, whereas in CSNX rats, the summed HX+HC responses were slightly greater than the HH response for MV.

Again, the initial responses to HX, HC and HH would be expected to be mediated by CB chemoafferents, and so it is important to establish what changes in MV during these challenges occurred within the first 60 sec in SHAM and CSNX rats. The changes in MV that occurred during the first 60 sec of HX, HC and HH are shown in the top panels of **Figure 12**. In SHAM rats (top left panel), the initial increases in MV elicited by HX, HC and HH were similar to one another. In CSNX rats (top right panel), the initial increases in MV elicited by HX, HC and HH were also similar to one another, and obviously smaller compared to the SHAM rats. Again, we wanted to determine whether the magnitude of the actual 60 sec MV response during HH was simply additive of that elicited by HX and HC alone. As seen in the middle panels of **Figure 12**, the actual 60 sec MV responses during HH in both SHAM and CSNX rats were substantially smaller than those of the computed HX+HC values. The bottom panels of **Figure 12** show the total MV responses that occurred during the first 60 sec of HX, HC and HH in SHAM rats (bottom left panel) and CSNX rats (bottom right panel). The initial increase in MV elicited by HH was less than would be expected by simple addition of the HX and HC responses in both SHAM and CSNX rats. Thus, as with (and because of) the fR responses, the concluding message of the 60 sec data is that there is a negative interaction between HX and HC signaling pathways during the first 60 sec of the HH gas challenge, and this is independent of the presence or absence of the CB-CSN complex.

DISCUSSION

Proper tissue oxygenation is a well-regulated process that is essential for the survival of air-breathing animal species. The CBs are crucial for detecting changes in arterial pO_2, pCO_2 and pH. A drop in arterial pO_2 (hypoxia) or a rise in pCO_2 (hypercapnia) stimulates the CBs, and the sensory information is then relayed to chemoafferents of the CSN, which in turn sends signals to the respiratory control regions in the brainstem to trigger downstream respiratory and cardiovascular reflexes (López-Barneo et al., 2008). Over the years research has shown developmental differences between hypoxia (HX) and hypercapnia (HC) ventilatory responses in neonatal and adult mammals (Eden and Hanson, 1987; Bureau et al., 1984; McCooke and Hanson 1985; Bonora et al., 1984; Elnazir et al., 1996; Easton et al., 1986; Vizek et al., 1987; Maxova and Vizek, 2001; Pokorski and Antosiewicz, 2010; Putnam et al., 2005; Stunden et al., 2001; Teppema and Dahan, 2010). In addition, studies have shown that CSNX blunts the HX and HC ventilatory responses in both adult and newborn animals (Baby et al., 2018; Olson et al., 1988; Sapru & Krieger, 1977; Mouradian Jr. et al., 2012; Cardenas and Zapata, 1983; Chalmers et al., 1967; Bureau et al., 1985; Fung et al., 1996; Lowry et al., 1999a; Lowry et al., 1999b; Suguihara et al., 1994; Teppema and Dahan, 2010). However, the effect of CSNX on HX (poikilocapnic), HC and hypoxia-hypercapnia (HH) ventilatory responses in juvenile (P25) rats remains unclear. Earlier studies have shown that Sprague Dawley rats at P25 are not sexually mature (Sengupta, 2013), but have fully mature peripheral and central neuronal systems (Dumas, 2005; Jiang et al., 2005; Tahayori and Koceja 2012; Sinclair et al., 2014; Lai et al., 2016). Thus, we used the P25 age group in this study to analyze the role of peripheral CB chemoafferents in the ventilatory performance of juvenile rats. We hypothesized that the peripheral chemosensory pathway in juvenile rats has a unique role in determining the interaction between O_2 and CO_2, and consequently, the response to hypoxic-hypercapnic gas challenges. Our present study provides compelling evidence supporting previous observations that peripheral and central chemoreflexes appear to be fully operational in Sprague Dawley P25 rats, because we found that the HX (10% O_2, 90% N_2) and HC (5% CO_2, 21% O_2, 90% N_2) gas challenges we exposed the P25 rats to elicited robust increases in fR, VT and MV in SHAM rats that resemble in many respects those observed in adult rats of various strains, including Sprague Dawley rats (Teppema and Dahan, 2010; May et al., 2013a,b).

HX and HC challenges are probes used to study and determine ventilatory signaling cascades elicited by decreases in arterial pO_2 or increases in arterial pCO_2, and the HH challenge is more physiological and clinically relevant in that it causes changes in pO_2 and pCO_2 blood-gas chemistry that resemble those elicited by central/obstructive sleep apneas (Smith et al., 1984, 2006, 2010, 2015; Blain et al., 2010; Forster and Smith, 2010; May et al., 2013a,b). The major objectives of this study were (1) to characterize the changes in fR, VT and MV elicited by HX, HC and HH in order to define and compare, how these challenges affect ventilation in juvenile rats; (2) to determine how ventilatory responses elicited by HX, HC and HH are dependent on the presence of CB chemoafferents; (3) to determine the interaction between HX and HC signaling pathways in SHAM rats in order to characterize whether these inputs are (a) simply additive (i.e., independent of one another), (b) synergistic (super-additive) or (c) inhibitory (hypo-additive); and (4) to determine whether the interaction between hypoxia and hypercapnia signaling pathways is under the influence of the CB-CSN chemoafferent complex (i.e. is the interaction disturbed in bilateral CSNX rats).

Our data shows that (1) the initial rate of rise (over the first 60 sec) in fR, VT and MV elicited by the HX, HC and HH gas challenges were shallower in CSNX rats than in SHAM rats, (2) the total increases in fR, VT and MV elicited by HX were diminished in CSNX rats compared to SHAM rats, (3) the total increases in fR, VT and MV elicited by HC in CSNX rats were not significantly different from those in SHAM rats, and (4) the total increases in fR, VT and MV elicited by HH observed in SHAM rats were markedly diminished in CSNX rats. These findings suggests and/or support previous evidence that (1) the activity of CB chemoafferents increases rapidly in response to changes in arterial blood pO_2 and pCO_2, thereby allowing more rapid ventilatory responses than central chemoreceptors (Fatemian et al., 2003; Smith et al., 2006), (2) the CB-CSN inputs to the cNTS play a discernible and important role in HX ventilatory signaling, while central chemoreceptors provide most of the steady-state response to elevations in blood pCO_2 (Prabhakar, 2000; López-Barneo et al., 2008; Teppema and Dahan, 2010; Guyenet and Bayliss, 2015; Guyenet et al., 2019), (3) the loss of CB-CSN complexes does not impair HC-induced enhancement of ventilatory performance, and (4) the VT response that occurs when HX and HC challenges are given simultaneously (i.e., the HH gas challenge), is indeed additive and this additivity is clearly dependent upon the CB-CSN input to the cNTS as well as direct activation of the retrotrapezoid nucleus (RTN) (Smith et al., 1984, 2006, 2010, 2015; Blain et al., 2010; Forster and Smith, 2010). However, it must be noted that the interactions of the cNTS and RTN with other key nuclei within the brainstem will most likely be of vital importance to the expression of the observed ventilatory responses. Moreover, the relative role of astrocytes and neurons in the brainstem with respect to hypercapnic signaling and the interaction of hypercapnic and hypoxic signaling pathways must also be considered (see Turovsky et al., 2016; Sheikhbahaei et al., 2016).

Furthermore, in SHAM rats, the total responses that occurred during the actual HH gas challenge were equal to simple addition of actual HX and HC values (i.e. HX+HC responses). The exception was that the initial increases in fR elicited by HH were less than would be predicted by addition of the HX and HC responses. Thus, our evidence for additivity is consistent with a previous study done in cats showing that when HX and HC stimuli are given together (i.e. HH stimuli), the interaction between central and peripheral chemoreceptors is additive in that one stimulus, HX or HC, simply adds to the other (Eldridge et al., 1981). However, our observation for additivity in P25 rats contradicts previous findings that (1) a positive (synergistic) interaction between HX-HC induced neuronal signaling in cats exists at the level of the CB (Eyzaguirre and Lewin, 1961), (2) in adult dogs (no relevant data in adult rats) that peripheral CB and central chemoreceptors are functionally inter-dependent such that activity and/or sensitivity of medullary chemoreceptors is determined by input from CB chemoreceptors and *vice versa* (Fencl et al., 1966; Smith et al., 1984, 2006, 2010, 2015; Dempsey, 2005; Blain et al., 2010; Forster and Smith, 2010). and (3) interaction between central and peripheral chemoreceptors in cats and rabbits is negative (i.e. the stimuli interact in a hypo-additive manner) (Eldridge, 1974; Kiwull-Schöne et al., 1976). In addition, it is evident from this study that fR, VT and

MV responses elicited by HH did not reach the maximum values (i.e. ceiling values) that the rats were capable of achieving throughout the entire 5 x 5 HX, HC or HH gas challenges (i.e., maximum values observed when the rats are grooming, exploring, or in a possible state of elevated vigilance in response to an unexpected external stimulus, such as a sound that the rats could hear, remembering that they were housed in plexi-glass chambers in a quiet room to control for external noise disturbances). Indeed, the markedly smaller than expected increase in fR elicited by HH gas challenge was not simply because the rats could not attain higher values. These findings therefore suggest that HH only recruits ventilatory pathways that have a set-point with respect to the maximum levels of responses that can be achieved.

It should be noted that the P25 rats used for these plethysmography studies were awake and quietly resting before and during the HX, HC and HH gas challenges. This is important to mention because the awake-sleep state of the rat can influence the magnitude of ventilatory responses to direct brain stimulation and in response to HX and HC gas challenges (Abbott et al., 2013). In particular, the ventilatory responses seen during awake and non-REM sleep stages are similar, but markedly blunted during REM sleep. In our studies, none of the rats were asleep at any stages of the study and as such we do not believe that the data was influenced by the awake-sleep status of the rats. In addition, it should be noted that robust ventilatory responses to poikilocapnic HX gas challenges were evident in the juvenile (P25) rats with prior bilateral CSNX. This finding differs strikingly to the much more dramatic loss of ventilatory responses to HX gas challenge in adult rats and mice (Bisgard et al., 1976; Sapru and Krieger, 1977; Smith and

Mills, 1980; Martin-Body et al., 1985, 1986; Olson et al., 1988; Roux et al., 2000; Baby et al., 2018). Whether the remaining ventilatory responses to HX challenge in P25 rats with bilateral CSNX is due to (1) low secondary O_2-sensing structures, such as aortic bodies or peripheral (e.g., glossopharyngeal) nerves (Eyzaguirre and Lewin, 1961); (2) the dynamic functional reorganization of central chemoreceptor reflex pathways (Roux et al., 2000); and/or (3) central O_2 sensors (Dillon and Waldrop, 1993; Sun and Reis, 1994), remains to be determined. Finally, our data suggesting that loss of CSN input to the brainstem markedly impairs the additivity of the hypoxic-hypercapnic ventilatory response, may have important implications for the development of disease processes whereby the loss of function of the CBs and/or CSN chemoafferents contributes to changes in ventilation in unexpected ways.

CONCLUSION

Overall, this novel data reveals that the CSN provides necessary input to respiratory control regions in the brainstem of juvenile (P25) rats in order to induce HX, HC and HH ventilatory responses. In addition, our data suggests that juvenile rats with prior CSNX possess a functional secondary means of low O_2 sensing, other than the CB-CSN complex, that also send signals to respiratory control centers of the brainstem to trigger downstream respiratory reflexes. Finally, our results reveal that additive between HX and HC stimuli is absent in rats with CSNX, therefore suggesting that loss of CSN chemoafferent signaling to the brainstem allows central CO_2 chemoreceptors to dominate the secondary low O_2 sensing input at the points of convergence between the central and peripheral chemoreceptor input.

In conclusion we have comprised a series of schematics describing how HX and HC signaling pathways increase ventilation in P25 male rats with the CSN intact and CSN transected (**Figure 13**). In SHAM rats (i.e., CSN intact), a hypoxic gas challenge (HX) triggers a *strong* (++) depolarization of chemosensitive glomus cells (GC) in the CB that causes the release of excitatory neurotransmitters. These neurotransmitters then activate chemosensory afferent fibers in the CSN, whose cell bodies are located in the petrosal ganglion (PG). The central projections of these chemoafferents signal neurons in the commissural nucleus tractus solitarius (cNTS), which relay downstream respiratory-related neuronal pathways in the central pattern generator (CPG) to increase ventilation (++) in an effort to restore blood gas homeostasis. A hypercapnic gas challenge (HC) triggers a *moderate* (+) depolarization, and release of neurotransmitters from chemosensitive GC in the CB. In addition, HC triggers a *moderate* (+) depolarization and release of neurotransmitters from central chemosensors in the retrotrapezoid nucleus (RTN). Both signaling pathways converge in the CPG to ultimately increase ventilation (++). A hypoxic-hypercapnic gas challenge (HH) triggers a larger (+++) depolarization compared to HX and release of neurotransmitters from chemosensitive GC in the CB. This is due to the positive interaction between hypoxia and hypercapnia when given together at the level of the CB. In addition, HH triggers a *moderate* (+) depolarization and release of neurotransmitters from central chemosensors in the RTN. Both signaling pathways converge in the central pattern generator (CPG) to ultimately cause an even greater increase ventilation (++++) that is additive of the HX and HC signaling pathways.

In CSNX rats (i.e., CSN transected), HX triggers secondary O_2-sensing structures (S–O_2–SS) that, for the sake of argument, may synapse within the cNTS. This S–O_2–SS input causes a *smaller* (+) activation of cNTS neurons compared to CSN chemoafferent input. The cNTS then relays downstream respiratory-related neuronal pathways in the CPG to increase ventilation (+). A HC triggers a *smaller* (+) depolarization, and release of neurotransmitters from S–O_2–SS, compared to CSN chemoafferent input. In addition, it triggers a *moderate* (+) depolarization from central chemosensors in the RTN. Both pathways converge in the CPG with the input from the RTN occluding the input from the S–O_2–SS. As such, the increase in ventilation (+) is mediated solely by the central chemoreceptors, and thus the additivity that is seen when the CSN is intact is lost. Finally, HH triggers a *smaller* (+) depolarization, and release of neurotransmitters from S–O_2–SS, compared to CSN chemoafferent input. In addition, it triggers a *moderate* (+) depolarization from central chemosensors in the RTN. Both pathways converge in the CPG with the input from the RTN occluding the input from the S–O_2–SS. As such, the increase in ventilation (+) is mediated solely by the central chemoreceptors, and thus the additivity that is seen when the CSN is intact is again lost. We understand that these schematics are rudimentary, and that there are many other plausible possibilities to explain our data for (1) the additivity of HX and HC signaling pathways in SHAM rats and (2) the loss of this additivity in CSNX rats. We are sure that this series of schematics will evolve as new data is collected.

Table 1. Baseline parameters in sham-operated rats and in those with bilateral transection of the carotid sinus nerve

Parameter	Group	HX Challenge	HC Challenge	HH Challenge
Number of rats	SHAM	14	12	15
	CSNX	14	12	15
Age, days post birth	SHAM	25	25	25
	CSNX	25	25	25
Body Weight, grams	SHAM	60 ± 3	68 ± 3	61 ± 2
	CSNX	62 ± 4	65 ± 2	59 ± 2
Values recorded throughout the acclimatization period				
Frequency, breaths/min	SHAM	123 ± 5	120 ± 2	125 ± 3
	CSNX	109 ± 3*	114 ± 2*	112 ± 3*
Tidal Volume, ml	SHAM	0.63 ± 0.02	0.68 ± 0.02	0.62 ± 0.01
	CSNX	0.55 ± 0.01*	0.59 ± 0.01*	0.56 ± 0.01*
Minute Ventilation, ml/min	SHAM	72 ± 4	79 ± 2	75 ± 1
	CSNX	59 ± 2*	65 ± 2*	60 ± 2*
Values recorded over the 90 sec prior to the gas challenges (i.e., Pre-values)				
Frequency, breaths/min	SHAM	113 ± 4	117 ± 2	119 ± 3
	CSNX	100 ± 2*	106 ± 2*	106 ± 3*
Tidal Volume, ml	SHAM	0.60 ± 0.02	0.66 ± 0.01	0.60 ± 0.01
	CSNX	0.55 ± 0.01*	0.58 ± 0.01*	0.55 ± 0.01*
Minute Ventilation, ml/min	SHAM	69 ± 3	79 ± 1	73 ± 2
	CSNX	56 ± 2*	63 ± 1*	59 ± 2*

SHAM, sham-operated rats. CSNX, rats with bilateral carotid sinus nerve transection. HX, hypoxic; HC, hypercapnic; HH, hypoxic-hypercapnic. Data are presented as mean $\pm$ SEM. *$P < 0.05$, CSNX *versus* SHAM.

Table 2. Maximal attainable values and maximal values recorded during the gas challenges

Parameter	Gas	Group	Pre-value	Maximum	Gas Challenge	Maximum delta	Gas Challenge delta
fR (bpm)	HX	SHAM	113 ± 4	355 ± 16	193 ± 10	+242 ± 16[†]	+80 ± 9[†,‡]
		CSNX	100 ± 2*	387 ± 9	164 ± 8	+287 ± 10[†]	+64 ± 8[†,‡]
	HC	SHAM	117 ± 2	351 ± 15	222 ± 18	+234 ± 17[†]	+105 ± 18[†,‡]
		CSNX	106 ± 2*	363 ± 15	190 ± 13	+257 ± 16[†]	+84 ± 14[†,‡]
	HH	SHAM	119 ± 3	343 ± 19	246 ± 11	+224 ± 19[†]	+126 ± 10[†,‡]
		CSNX	106 ± 3*	377 ± 14	211 ± 16	+271 ± 14[†]	+105 ± 15[†,‡]
VT (ml)	HX	SHAM	0.60 ± 0.02	1.12 ± 0.05	0.97 ± 0.04	+0.53 ± 0.04[†]	+0.37 ± 0.03[†,‡]
		CSNX	0.55 ± 0.01*	1.09 ± 0.05	0.82 ± 0.03	+0.54 ± 0.05[†]	+0.27 ± 0.03[†,‡]
	HC	SHAM	0.66 ± 0.01	1.14 ± 0.04	1.03 ± 0.04	+0.48 ± 0.03[†]	+0.37 ± 0.03[†,‡]
		CSNX	0.58 ± 0.01*	1.00 ± 0.02	0.88 ± 0.03	+0.42 ± 0.02[†]	+0.30 ± 0.02[†,‡]
	HH	SHAM	0.60± 0.01	1.35 ± 0.04	1.22 ± 0.04	+0.74 ± 0.04[†]	+0.62 ± 0.03[†,‡]
		CSNX	0.55 ± 0.01*	1.00 ± 0.03	0.87 ± 0.02	0.45 ± 0.03[†]	+0.32 ± 0.02[†,‡]
MV (ml/min)	HX	SHAM	69 ± 3	222 ± 13	154 ± 6	+153 ± 11[†]	+84 ± 5[†,‡]
		CSNX	56 ± 2*	234 ± 13	103 ± 5	+178 ± 12[†]	+46 ± 5[†,‡]
	HC	SHAM	79 ± 1	238 ± 12	178 ± 14	+159 ± 11[†]	+99 ± 13[†,‡]
		CSNX	63 ± 1*	203 ± 13	133 ± 6	+140 ± 13[†]	+70 ± 5[†,‡]
	HH	SHAM	73 ± 1	262 ± 5	240 ± 4	+190 ± 6[†]	+168 ± 4[†,‡]
		CSNX	59 ± 2*	198 ± 6	145 ± 7	+138 ± 6[†]	+85 ± 6[†,‡]

SHAM, sham-operated rats. CSNX, rats with bilateral carotid sinus nerve transection. fR (bpm), frequency of breathing (breaths/min); VT, tidal volume; MV, minute ventilation. HX, hypoxic gas challenge; HC, hypercapnic gas challenge; HH, hypoxic-hypercapnic gas challenge. Data are presented as mean ± SEM. *P < 0.05, pre-values, CSNX *versus* SHAM. [†]P < 0.05, delta responses, significant changes from pre-values. [‡]P < 0.05, Hypoxia delta response *versus* maximal attainable delta response.

Table 3. Arithmetic differences between the responses observed in sham-operated rats and those with bilateral transection of the carotid sinus nerve

	Frequency (breaths/min)		Tidal Volume (ml)		Minute Ventilation (ml/min)	
	Delta ACTUAL	Delta HX+HC	Delta ACTUAL	Delta HX+HC	Delta ACTUAL	Delta HX+HC
HX 1	-419		-2.01		-667	
HC 1	-451		-0.43		-635	
HH 1	-554	-870	-4.65	-2.44	-1448	-1302

The data are presented as the mean of the differences in frequency of breathing (breaths/min), tidal volume (ml) and minute ventilation (ml/min) between the bilateral carotid sinus nerve transected rats and the sham-operated rats. HX, hypoxic gas challenge; HC, hypercapnic gas challenge; HH, hypoxic-hypercapnic gas challenge.

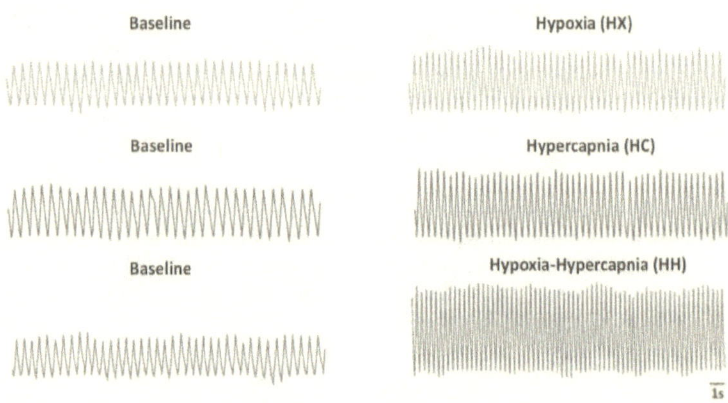

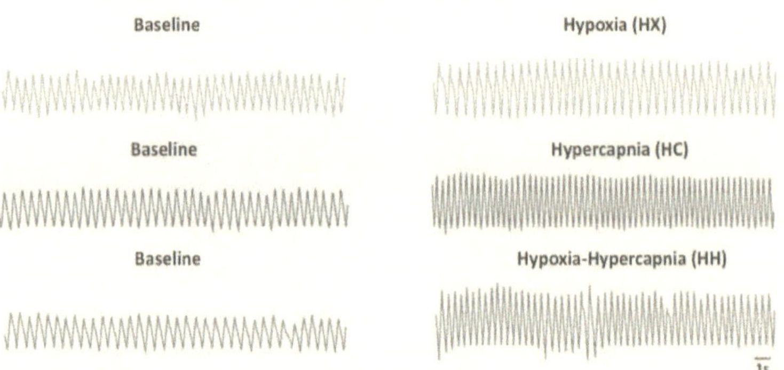

Figure 3. Representative plethysmography traces from sham-operated (SHAM) P25 rats (**Top panel**) and bilateral carotid sinus nerve transection (CSNX) (**Bottom panel**) P25 rats recorded at baseline and hypoxic (HX; 10% O_2, 90% N_2), hypercapnic (HC; 5% CO_2, 21% O_2, 74% N_2), or hypoxic-hypercapnic (HH; 5% CO_2, 10% O_2, 85% N_2) gas challenges.

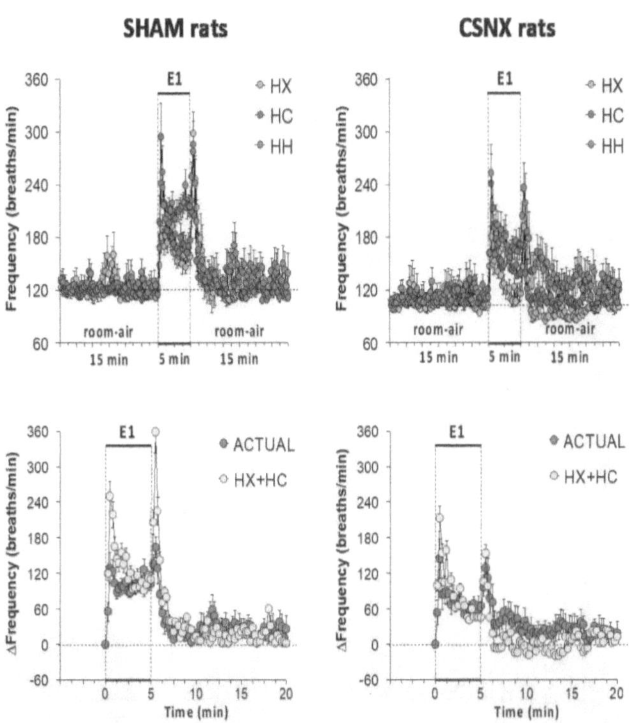

Figure 4. Top panels: Frequency of breathing (fR) values averaged every 15 sec in freely-moving sham-operated (SHAM) rats and in rats with bilateral carotid sinus nerve transection (CSNX) during exposure to a single episode (E1) of hypoxic (HX; 10% O_2, 90% N_2), hypercapnic (HC; 5% CO_2, 21% O_2, 74% N_2), or hypoxic-hypercapnic (HH; 5% CO_2, 10% O_2, 85% N_2) gas challenge of 5 min in duration, followed by 15 min of room-air. The data are presented as mean $\pm$ SEM. **Bottom panels:** Changes in fR in SHAM and CSNX rats during the actual HH gas challenge compared to addition of HX+HC values. For the actual HH values, data are presented as mean $\pm$ SEM and for the HX+HC values, data are presented as the mean $\pm$ SEM

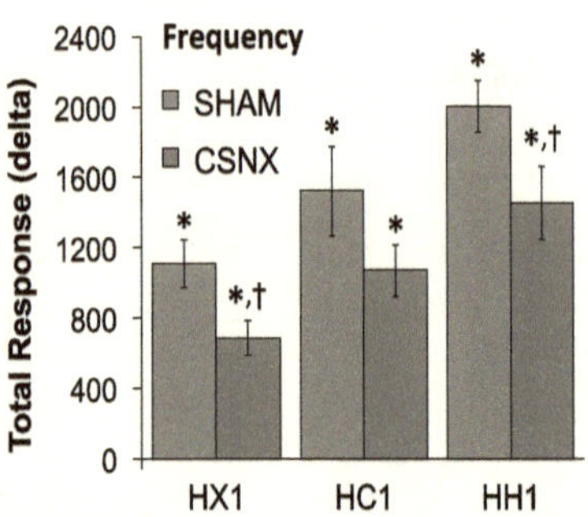

SHAM rats   **CSNX rats**

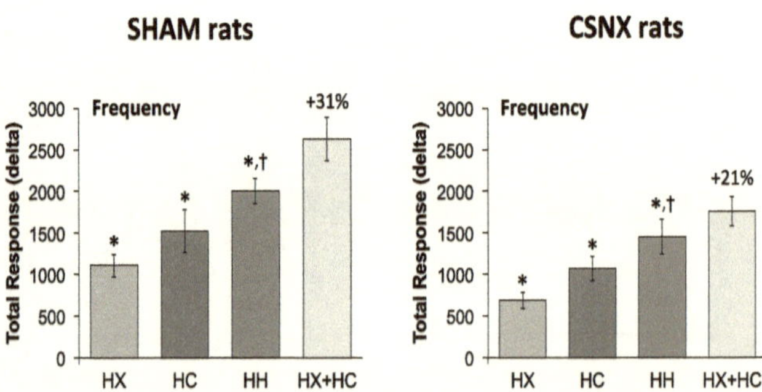

Figure 5. Top panels: Total arithmetic changes in frequency of breathing (fR) that occurred during exposure to a single hypoxic (HX1, 10% O_2, 90% N_2), hypercapnic (HC1, 5% CO_2, 21% O_2, 74% N_2) or hypoxic-hypercapnic (HH1, 5% CO_2, 10% O_2, 85% N_2) gas challenge in SHAM and CSNX rats. The data are presented as mean ± SEM. *P < 0.05, significant response from Pre-value. [†]P < 0.05, CSNX *versus* SHAM. **Bottom panels:** Total arithmetic changes in fR values in SHAM (left panel) and CSNX (right panel) rats during exposure to the HX, HC or HH gas challenges. The sum of HX and HC values (HX+HC) expressed as mean ± SEM (10% of mean) are shown. The figures in parentheses above the HX+HC columns are the percentage (%) differences between actual HH values and HX+HC values. *P < 0.05, significant response from Pre-value. [†]P < 0.05, HH *versus* HX. [‡]P < 0.05, HH *versus* HC.

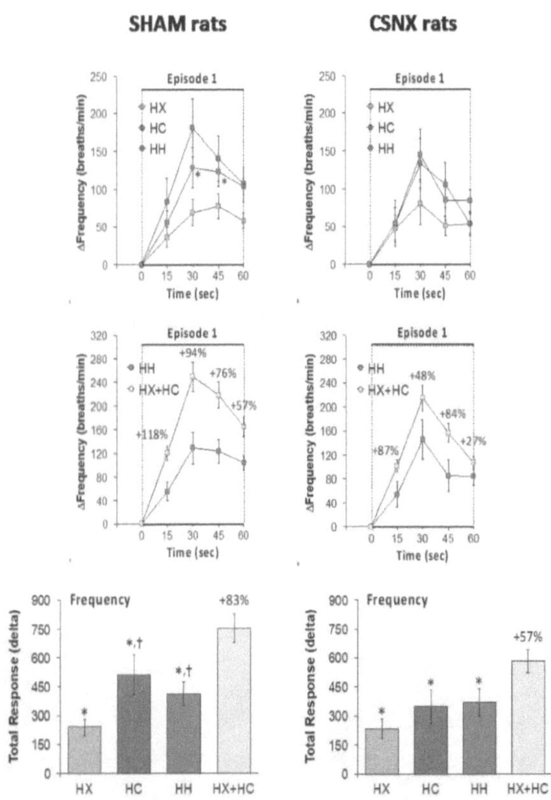

Figure 6. Top panels: Arithmetic changes in frequency of breathing (fR) in freely-moving sham-operated (SHAM) rats and in rats with bilateral carotid sinus nerve transection (CSNX) during the first 60 sec of exposure to a single hypoxic (HX, 10% O_2, 90% N_2), hypercapnic (HC, 5% CO_2, 21% O_2, 74% N_2), or hypoxic-hypercapnic (HH, 5% CO_2, 10% O_2, 85% N_2) gas challenge (Episode 1 (E1)). Data are presented as mean ± SEM. *P < 0.05, HH versus HX. [†]P < 0.05, HH *versus* HC. **Middle panels:** Arithmetic changes in fR in SHAM and CSNX rats during the first 60 sec of the actual hypoxic-hypercapnic (HH) gas challenge values compared to the addition of the HX and HC (HX+HC) gas challenge values. For the actual HH values, data are shown as mean ± SEM and for HX+HC values, data are presented as the mean ± SEM (10% of mean). **Bottom panels:** Total arithmetic changes in fR in SHAM and CSNX rats during the first 60 sec of exposure to HX, HC or HH gas challenges. The data are presented as mean ± SEM. The sum of the HX and HC values (HX+HC), expressed as mean ± SEM (10% of mean), are also shown. The numbers above the HX+HC columns are the percentage (%) differences between the HX+HC values and actual HH values. *P < 0.05, significant response from Pre-value. [†]P < 0.05, HH *versus* HX. [‡]P < 0.05, HH

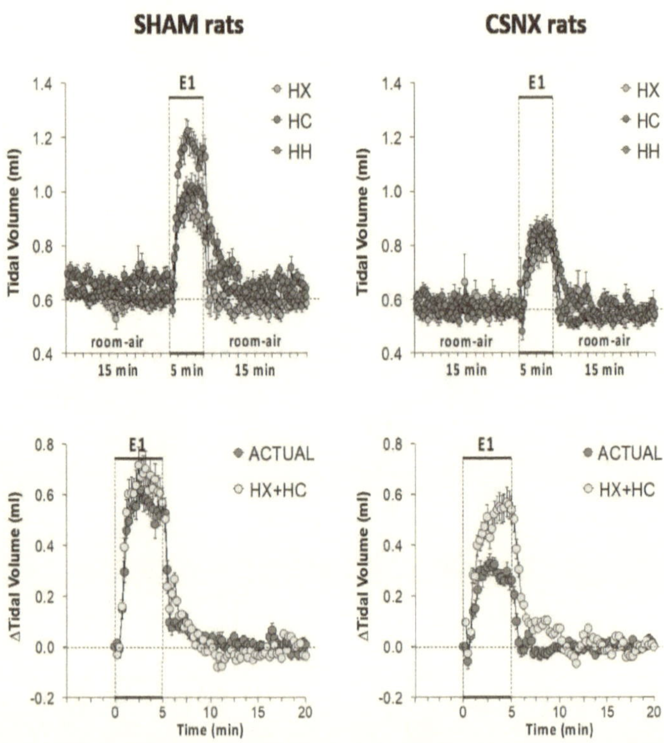

Figure 7. Top panels: Tidal volume (VT) values averaged every 15 sec in freely-moving sham-operated (SHAM) rats and in rats with bilateral carotid sinus nerve transection (CSNX) during exposure to a single episode (E1) of hypoxic (HX; 10% O_2, 90% N_2), hypercapnic (HC; 5% CO_2, 21% O_2, 74% N_2), or hypoxic-hypercapnic (HH; 5% CO_2, 10% O_2, 85% N_2) gas challenge of 5 min in duration, followed by 15 min of room-air. The data are presented as mean ± SEM. **Bottom panels:** Changes in VT in SHAM and CSNX rats during the actual HH gas challenge compared to addition of HX+HC values. For the actual HH values, data are presented as mean ± SEM and for the HX+HC values, data are presented as the mean ± SEM (10% of mean).

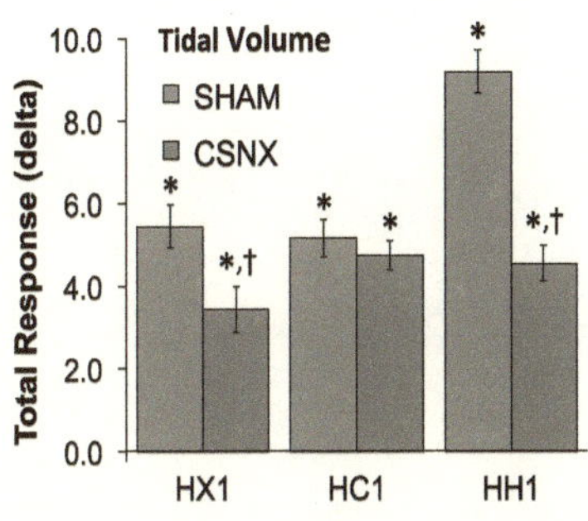

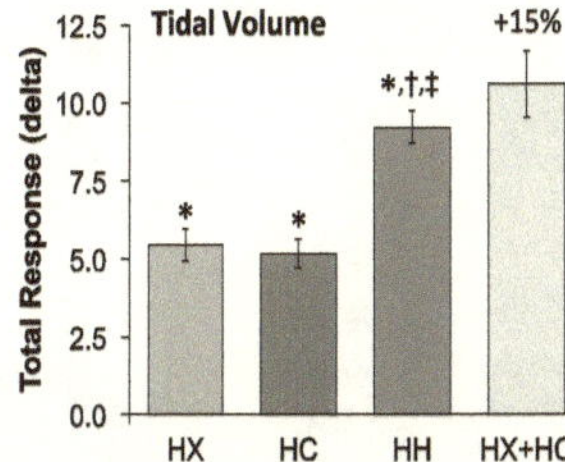

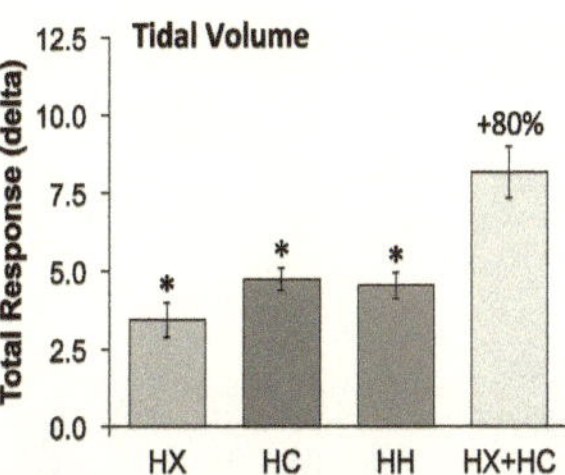

Figure 8. Top panels: Total arithmetic changes in tidal volume (VT) that occurred during exposure to a single hypoxic (HX1, 10% O_2, 90% N_2), hypercapnic (HC1, 5% CO_2, 21% O_2, 74% N_2) or hypoxic-hypercapnic (HH1, 5% CO_2, 10% O_2, 85% N_2) gas challenge in SHAM and CSNX rats. The data are presented as mean ± SEM. *$P < 0.05$, significant response from Pre-value. [†]$P < 0.05$, CSNX *versus* SHAM. **Bottom panels:** Total arithmetic changes in VT values in SHAM (left panel) and CSNX (right panel) rats during exposure to HX, HC or HH gas challenges. The sum of the HX and HC values (HX+HC) expressed as mean ± SEM (10% of mean) are shown. The figures in parentheses above the HX+HC columns are the percentage (%) differences between actual HH values and HX+HC values. *$P < 0.05$, significant response from Pre-value. [†]$P < 0.05$, HH *versus* HX. [‡]$P < 0.05$, HH *versus* HC.

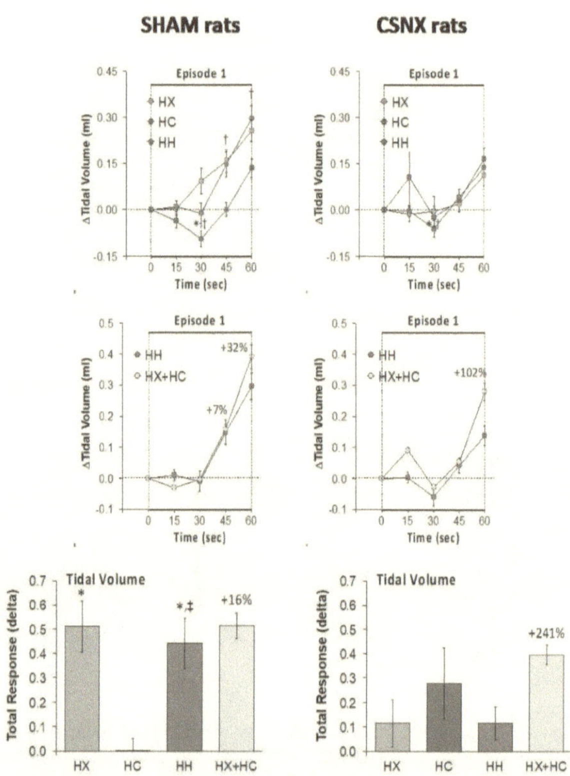

Figure 9. Top panels: Arithmetic changes in tidal volume (VT) in freely-moving sham-operated (SHAM) rats and in rats with bilateral carotid sinus nerve transection (CSNX) during the first 60 sec of exposure to a single hypoxic (HX, 10% O_2, 90% N_2), hypercapnic (HC, 5% CO_2, 21% O_2, 74% N_2), or hypoxic-hypercapnic (HH, 5% CO_2, 10% O_2, 85% N_2) gas challenge (Episode 1 (E1)).

Data are presented as mean ± SEM. *P < 0.05, HH versus HX. [†]P < 0.05, HH *versus* HC. **Middle panels:** Arithmetic changes in VT in SHAM and CSNX rats during the first 60 sec of the actual hypoxic-hypercapnic (HH) gas challenge values compared to addition of the HX and HC (HX+HC) gas challenge values. For the actual HH values, data are shown as mean ± SEM and for HX+HC values, data are presented as the mean ± SEM (10% of mean). **Bottom panels:** Total arithmetic changes in VT in SHAM and CSNX rats during the first 60 sec of exposure to HX, HC or HH gas challenges. Data are presented as mean ± SEM. The sum of HX and HC values (HX+HC), expressed as mean ± SEM (10% of mean), are shown. The numbers above the HX+HC columns are the percentage (%) differences between the HX+HC values and actual HH values.

*P < 0.05, significant response from Pre-value. [†]P < 0.05, HH *versus* HX. [‡]P < 0.05, HH *versus*

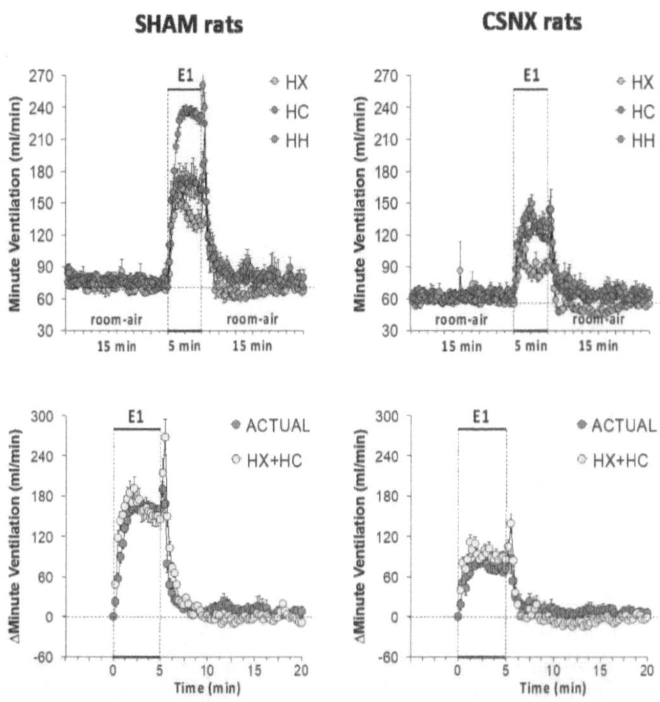

Figure 10. Top panels: Minute ventilation (MV) values averaged every 15 sec in freely-moving sham-operated (SHAM) rats and in rats with bilateral carotid sinus nerve transection (CSNX) during exposure to a single episode (E1) of hypoxic (HX; 10% O_2, 90% N_2), hypercapnic (HC; 5% CO_2, 21% O_2, 74% N_2), or hypoxic-hypercapnic (HH; 5% CO_2, 10% O_2, 85% N_2) gas challenge of 5 min in duration, followed by 15 min of room-air. The data are presented as mean ± SEM. **Bottom panels:** Changes in MV in SHAM and CSNX rats during the actual HH gas challenge compared to addition of HX+HC values. For the actual HH values, data are presented as mean ± SEM and for the HX+HC values, data are presented as the mean ± SEM (10% of mean).

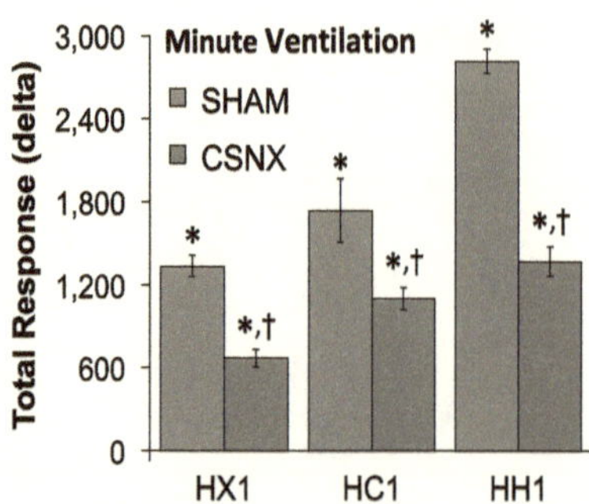

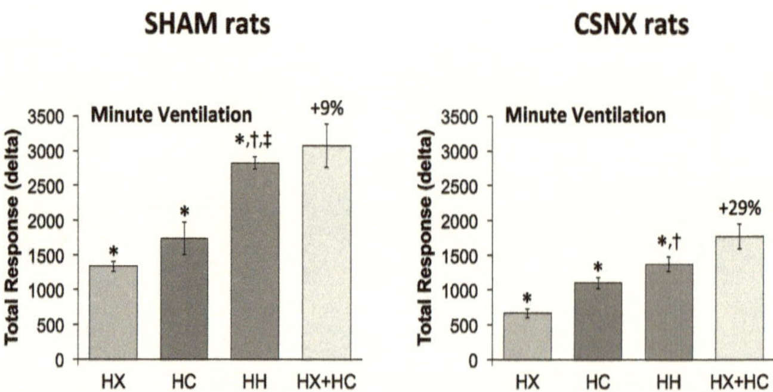

Figure 11. Top panels: Total arithmetic changes in minute ventilation (MV) that occurred during exposure to a single hypoxic (HX1, 10% O_2, 90% N_2), hypercapnic (HC1, 5% CO_2, 21% O_2, 74% N_2) or hypoxic-hypercapnic (HH1, 5% CO_2, 10% O_2, 85% N_2) gas challenge in SHAM and CSNX rats. The data are presented as mean ± SEM. *$P < 0.05$, significant response from Pre-value. †$P < 0.05$, CSNX *versus* SHAM. **Bottom panels:** Total arithmetic changes in MV in SHAM (left panel) and CSNX (right panel) rats during exposure to the HX, HC or HH gas challenges. The sum of HX and HC values (HX+HC) expressed as mean ± SEM (10% of mean) are shown. The figures in parentheses above the HX+HC columns are the percentage (%) differences between actual HH values and HX+HC values. *$P < 0.05$, significant response from Pre-value. †$P < 0.05$, HH *versus* HX. ‡$P < 0.05$, HH *versus* HC.

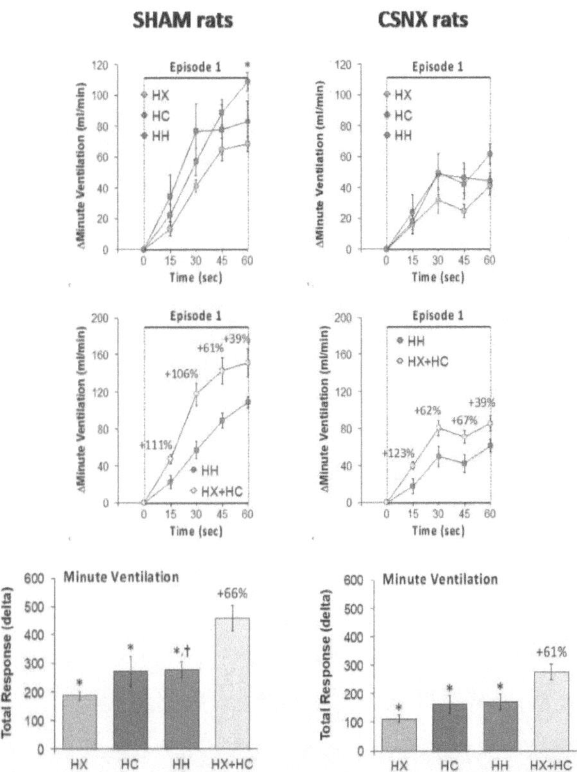

Figure 12. Top panels: Arithmetic changes in minute ventilation (MV) in freely-moving sham-operated (SHAM) rats and in rats with bilateral carotid sinus nerve transection (CSNX) during the first 60 sec of exposure to a single hypoxic (HX, 10% O_2, 90% N_2), hypercapnic (HC, 5% CO_2, 21% O_2, 74% N_2), or hypoxic-hypercapnic (HH, 5% CO_2, 10% O_2, 85% N_2) gas challenge (Episode 1 (E1)). Data are presented as mean $\pm$ SEM. *$P < 0.05$, HH versus HX. [†]$P < 0.05$, HH *versus* HC. **Middle panels:** Arithmetic changes in MV in SHAM and CSNX rats during the first 60 sec of the actual hypoxic-hypercapnic (HH) gas challenge values compared to addition of the HX and HC (HX+HC) gas challenge values. For the actual HH values, data are shown as mean $\pm$ SEM and for HX+HC values, data are presented as the mean $\pm$ SEM (10% of mean). **Bottom panels:** Total arithmetic changes in MV in SHAM and CSNX rats during the first 60 sec of exposure to HX, HC or HH gas challenges. The data are presented as mean $\pm$ SEM. The sum of the HX and HC values (HX+HC), expressed as mean $\pm$ SEM (10% of mean), are also shown. The numbers above the HX+HC columns are the percentage (%) differences between the HX+HC values and actual HH values. *$P < 0.05$, significant response from Pre-value. [†]$P < 0.05$, HH *versus* HX. [‡]$P < 0.05$, HH *versus* HC.

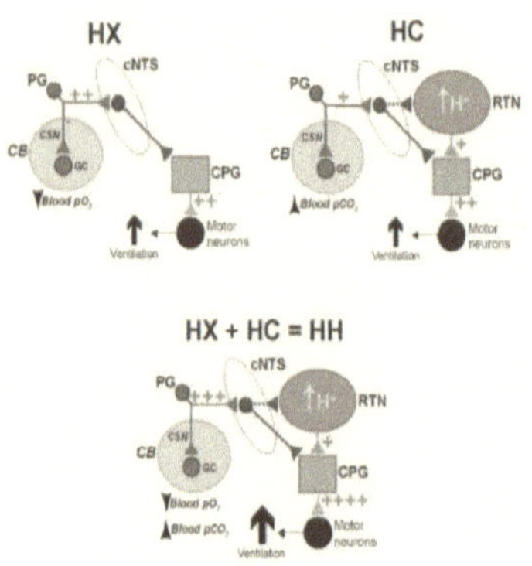

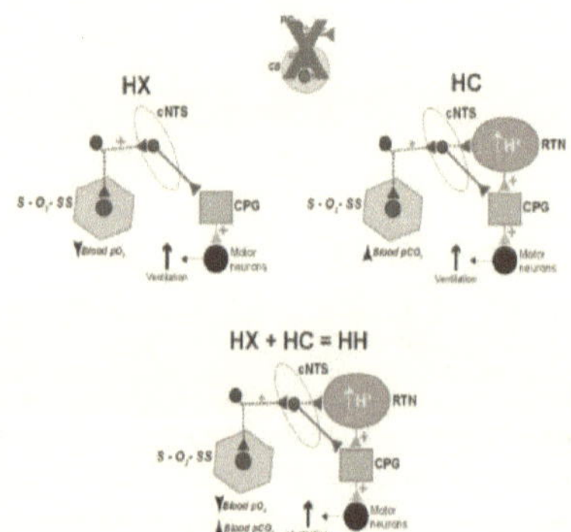

Figure 13. Schematic describing how activation of hypoxic and hypercapnic signaling pathways increase ventilation in P25 male SD rats in a simple additive manner and how this additivity is lost with prior CSN transection. HX, hypoxic gas challenge. HC, hypercapnic gas challenge. HH, hypoxic-hypercapnic gas challenge. CB, carotid body. GC, glomus cell. CSN, carotid sinus chemoafferent nerve. PG, petrosal ganglion cell body of the chemoafferent nerve. cNTS, commissural nucleus tractus solitarius. RTN, retrotrapezoid nucleus. CPG, central pattern generator.

CHAPTER III: LOSS OF CERVICAL SYMPATHETIC CHAIN INPUT TO THE SUPERIOR CERVICAL GANGLIA AFFECTS THE VENTILATORY RESPONSES TO HYPOXIC CHALLENGE IN FREELY-MOVING C57BL6 MICE

INTRODUCTION

Preganglionic sympathetic neurons emanating from the thoracic spinal cord (T1-T4) course in the left and right cervical sympathetic chains (CSC) and terminate on the cell bodies of postganglionic sympathetic neurons in the ipsilateral superior cervical ganglia (SCG) (Rando et al., 1981; Tang et al., 1995a,b,c; Llewellyn-Smith et al., 1998). The majority of postganglionic cells leave the SCG via the internal and external carotid nerves (Bowers and Zigmond, 1979; Buller and Bolter, 1997; Asomoto, 2004; Savastano et al., 2010), with the ganglioglomerular nerve (GGN) branching from the external carotid nerve to innervate primary glomus cells, chemoafferent nerve terminals, and vasculature within the carotid body (Biscoe and Purves, 1967; Zapata et al., 1969; Bowers and Zigmond, 1979; McDonald and Mitchell, 1981; McDonald, 1983a; Verna et al., 1984; Torrealba and

Claps, 1988; Ichikawa, 2002; Asamoto, 2004; Savastano et al., 2010), and baroafferent nerve terminals within the carotid sinus (Floyd and Neil, 1952; Rees, 1967; Bolter and Ledsome, 1976; Felder et al., 1983; Buller and Bolter, 1993).

The projections of the SCG to a series of inter-related structures, such as the carotid body and carotid sinus, upper airway and tongue (Flett and Bell, 1991; Wang and Chiou, 2004; Oh et al., 2006; Kummer et al., 1992; Hisa et al., 1999; O'Halloran et al, 1996, 1998), and nuclei within the hypothalamus and brainstem, including the nucleus tractus solitarius (nTS) (Cardinali et al., 1981a,b, 1982; Gallardo et al., 1984; Saavedra, 1985; Wiberg and Widenfalk, 1993; Westerhaus and Loewy, 1999; Esquifino et al., 2004; Hughes-Davis et al., 2005; Mathew, 2007), provide evidence that the SCG is a vital integrative structure regulating cardiorespiratory function. Previous studies have shown that electrical stimulation of the CSC decreases arterial blood pressure and upper airway resistance in rats (O'Halloran et al, 1996, 1998). Moreover, there is substantial (and conflicting) evidence as to the ability of sympathetic innervation to the carotid body to influence resting activity of glomus cells and chemoafferents within the carotid sinus nerve (CSN) and modulate changes in activity during hypoxic gas challenge (Prabhakar, 1994). For instance, CSC and GGN activity increases during hypoxic challenge, suggesting the release of neurotransmitters, such as norepinephrine, dopamine and neuropeptide Y (Matsumoto et al., 1986, 1987; Lahiri et al., 1986, Yokoyama et al., 2015), and conversely activation of the GGN decreases the hypoxic response of chemosensors within the cat carotid body (McQueen et el., 1989). A variety of responses have also been reported upon application of neurotransmitters released by CSC and GGN nerve terminals (e.g.

norepinephrine, dopamine and neuropeptide Y) to *in vivo* and *in vitro* carotid body preparations, including (1) a biphasic pattern consisting of initial brief bursts in carotid sinus nerve (CSN) activity and then long-lasting inhibition (Bisgard et al., 1979), (2) a biphasic pattern consisting of initial brief decreases in CSN activity and then long-lasting excitation (Matsumoto et al., 1981), (3) indirect excitation of carotid body glomus cells via constriction of arteriolar blood flow in the carotid body (Potter et al., 1987; Yokoyama et al., 2015), (4) direct activation of glomus cells and/or chemosensory afferents (Matsumoto et al., 1980; Lahiri et al., 1981; Milsom et al., 1983; Heinert et al., 1995; Pang et al., 1999), and (5) direct inhibitory action of glomus cells and/or chemosensory afferents (Zapata et al., 1969; Zapata, 1975; Llados and Zapata, 1978. Mills, et al., 1978; Folgering et al., 1982; Kou et al., 1991; Pizarro et al., 1992; Bisgard et al., 1993; Prabhakar et al., 1993; Ryan et al., 1995; Almaraz, et al., 1997; Overholt and Prabhakar, 1999).

Wild-type and genetically engineered mice are widely used to understand the mechanisms by which hypoxic challenges elicit carotid body-*dependent* and -*independent* ventilatory responses (He et al., 2000, 2002, 2003; Kline and Prabhakar, 2000; Pérez-García et al., 2004; Kline et al., 2005; Pichard et al., 2015; Gao et al., 2017; Wang et al., 2017; Ortega-Sáenz et al., 2018; Peng et al., 2018; Prabhakar et al, 2018). The morphology, neurophysiology and neuropharmacology of the mouse CSC-SCG has received considerable investigation over the years (Black et al., 1972; Yokota and Yamauchi, 1974; Banks and Walter, 1975; Inoue K, 1975; Lewis and Burton, 1977; Forehand, 1985; Kidd and Heath, 1988; Gibbins, 1991; Kasa et al., 1991; Little and Heath, 1994; Jobling and Gibbins, 1999; El-Fadaly and Kummer, 2003; David et al., 2010;

Cadaveira-Mosquera et al., 2012; Pashai et al., 2012; Alberola-Die et al., 2013; Liu et al., 2014; Martinez-Pinna et al., 2018; Mitsuoka et al., 2018; Feldman-Goriachnik and Hanani, 2019; Simeone et al., 2019; Rivas-Ramírez et al., 2020). Mouse tissues that receive postganglionic projections from the SCG have also been heavily investigated (Krieger, et al., 1976; García et al., 1988; Kawaja and Crutcher, 1997; Maklad et al., 2001; Pankevich et al., 2003a,b; Ivanusic et al., 2013; Karlsen et al., 2013; Lindborg et al., 2018; Ziegler et al., 2018; Teshima et al., 2019). Nonetheless, there is no direct evidence that the CSC-SCG complex innervates the carotid bodies of mice, and this is likely on the basis of the presence of sympathetic (i.e., tyrosine-hydroxylase-positive) nerve terminals and adrenergic receptors in these structures (Prieto-Lloret et al., 2007; Roux et al., 2008; Kåhlin et al., 2010; Chai et al., 2011).

Patients with chronic T1-T4 spinal cord injury present with cardiorespiratory disturbances consistent with diminished activity of the CSC-SCG complex (DiMarco et al., 2009; Heutink et al., 2014; Sankari et al., 2014, 2019; Berlowitz et al., 2016; Hachmann et al., 2017; Shin et al., 2019) including, *enhanced* peripheral (carotid body) chemoreflex sensitivity, which is a primary cause of sleep-disordered breathing in these subjects (Tester et al., 2014; Bascom et al., 2016). In contrast, studies in rats have found that mid-thoracic spinal cord injury is associated with enhanced cardiac sympathetic activity and cardiac sympathetic hyperinnervation that increases the susceptibility to life-threatening arrhythmias (Rodenbaugh et al., 2003; Collins et al., 2006; Lujan and DiCarlo 2007; Lujan et al., 2009, 2010, 2012, 2014). Studies in rats have also confirmed that hypoactivity of the CSC greatly enhances the likelihood of stroke in hypertensive models (Sadoshima et al., 1981; Sadoshima et al., 1983; Werber and Heistad, 1984).

The effects of bilateral transection of the CSC (CSCX) or bilateral removal of the SCG (SCGX) have been investigated on a variety of physiological functions/variables in the mouse (Krieger, et al., 1976; García et al., 1988; Kawaja and Crutcher, 1997; Pankevich et al., 2003a,b; Karlsen et al., 2013; Lindborg et al., 2018; Ziegler et al., 2018). However, to our knowledge, no studies, in any species, have determined the roles of the CSC-SCG complex on ventilatory functions and responses to hypoxic challenges *in vivo*. The aim of this study was to determine the effects of sham operation (SHAM) and bilateral CSCX (performed 4 days before testing) on resting ventilatory parameters in freely-moving adult male C57BL6 mice, and on their ventilatory responses to hypoxic gas challenge (HXC) using whole-body plethysmography as described previously (Gaston et al., 2014; Getsy et al., 2014; Palmer et al., 2013a,b; Palmer et al., 2015).

The C57BL6 mouse has proven to be an invaluable murine model in which to investigate the physiological systems involved in ventilatory control processes (Tankersey et al., 1994, 2000) and this mouse is used widely to generate gene knock-out mutants to investigate the molecular mechanisms underlying the responses of mice to hypoxic and/or hypercapnic gas challenges (Kline et al., 2000, 2005; Palmer et al., 2013a). The ventilatory responses that will be described in the SHAM mice (during and following the HXC) were consistent with previous studies from our laboratory (Palmer et al., 2013a,b; 2015; Gaston et al., 2014, Getsy et al., 2014). Our Getsy et al (2014) manuscript provided a detailed set of analyses of the ventilatory responses that occur in C57BL6, Swiss Webster and B6AF1 mice before, during and after a brief HXC and found, for example, that (1) the HVR in C57BL6 mice consists of an initial increase in frequency of breathing (Freq) followed by substantial decline (roll-off) toward pre-HXC values whereas the Freq responses in Swiss Webster and B6AF1 mice were robust with minimal roll-off, and (2) the post-HXC (return to room-air) responses consisted of a rapid and sustained rise in Freq in C57BL6 mice, a sustained rise in the B6AF1 mice, a form of short-term potentiation (STP) (Powell et al., 1998), and a gradual return to pre-hypoxia levels in the Swiss-Webster mice.

The ways and mechanisms by which activation of sympathetic nerves affects carotid body function under normoxic and hypoxic conditions have received considerable attention (Overholt and Prabhakar, 1999). On the basis of compelling evidence that sympathetic nerve terminals innervate glomus cells and the microvasculature within the carotid bodies (Vázquez-Nin et al., 1978; McDonald, 1983a; Verna et al., 1994) and that norepinephrine is the major neurotransmitter released by these terminals (Almaraz et al., 1997), the roles of this catecholamine and α_1-, α_2- and β-adrenoceptors and dopamine receptors have been studied. It is apparent that activation of sympathetic nerves can induce a multiplicity of effects within the carotid body. First, activation of the sympathetic nerves indirectly activate glomus cells by constricting arterioles (by activation α_1-adrenoceptors and dopamine receptors) in the carotid body, effectively resulting in a hypoxic environment for glomus cells (Llados and Zapata, 1978; Majcherczyk et al., 1980; Matsumoto et al., 1981; Yokoyama et al., 2015). In addition, co-release of neuropeptide Y from sympathetic nerve terminals also reduces blood flow within the carotid bodies

(Potter and McCloskey, 1987). Moreover, activation of β-adrenoceptors and dopamine receptors on glomus cells directly activates these cells (Eldridge and Gill-Kumar, 1980; Lahiri et al., 1981; Gonsalves et al., 1984). On the other hand, intra-carotid artery infusions of norepinephrine depress resting chemoreceptor activity and also attenuate hypoxic excitation of the carotid body, effects mediated by α_2-adrenoceptors (Kou et al., 1991; Pizarro et al., 1992; Prabhakar et al., 1993; Almaraz et al., 1997). Moreover, endogenous norepinephrine indirectly inhibits carotid body chemoafferent activity (Overholt and Prabhakar, 1999) via inhibition of glomus cell activity (presumably the release of excitatory neurotransmitters) by decreasing the magnitude and rate of activation of the macroscopic Ca^{2+} currents (Overholt and Prabhakar 1999). As such, it is open to question how the absence of functional sympathetic input to the carotid bodies and other structures controlling ventilatory processes would affect baseline parameters and the responses to HXC.

As mentioned, it has been established that HXC increases neural activity in the CSC and GGN (Matsumoto et al., 1986, 1987; Lahiri et al., 1986). As such, it is proposed that HXC activates a carotid sinus (chemoafferent) nerve-brainstem-descending spinal cord pathway that increases preganglionic sympathetic nerve activity, which in turn activates postganglionic neurons in the SCG, including those projecting to the carotid body via the ECN-GGN pathway. It should be noted that there is compelling evidence that SCG cells are not directly responsive to hypoxia (Rigual et al., 1999; Nunes et al., 2012; Buckler and Turner, 2013; Gao et al., 2019; Bernardini et al., 2020), although there is equally compelling evidence that SCG cells (and in particular small intensely fluorescent cells) are hypoxia-sensitive (Hanson et al., 1986; Brokaw and Hansen, 1987; Dinger et al., 1993; Strosznajder et al., 1997; Nunes et al., 2010, 2012).

The principal finding of this and our companion study (Getsy et al., 2020) is that transection of the CSC is not equivalent to bilateral removal of the SCG, suggesting multiple effects of hypoxia on neural signaling within the CSC-SCG pathway.

RESULTS

RESTING PARAMETERS

A summary of the mouse descriptors and resting ventilatory parameters is provided in **Table 4**. There were 11 mice in the SHAM group and 12 mice in the CSCX group. The ages and body weights of the two groups were similar to one another (P > 0.05, for both comparisons). Accordingly, no corrections for body weights were applied to the ventilatory data pertaining to volumes (e.g., tidal volume and peak inspiratory and expiratory flows). There were no between-group differences for any recorded or calculated ventilatory parameter (P > 0.05, for all comparisons). In the figures to follow, the left panel of each row will show the actual values recorded before, during the 5 min hypoxic (10% O_2, 90% N_2) challenge and upon to room-air. The middle panels of each row will show the arithmetic change recorded over the first 60 sec of hypoxic exposure. The right panels of each row will show the total response (% change from pre) during three epochs of the 5 min (300 sec) hypoxic challenge, namely 0-105 sec, 106-300 sec and 0-300 sec. These epochs were those which best represented the dramatic differences between SHAM and SCGX mice described in our companion manuscript (Getsy et al.,

2021).

HYPOXIC CHALLENGES - FREQUENCY OF BREATHING, TIDAL VOLUME AND MINUTE VENTILATION

Examples of respiratory waveforms during various stages of the experiment in a SHAM mouse and a CSCX mouse are shown in **Figure 14**. A time bar (1.25 sec) is shown in the bottom left of the figure. Resting Freq (pre-HXC values) was similar in both mice. The initial HXC response (HXC at 15 sec) was far greater in the CSCX mouse (166 to 360 breaths/min = 194 breaths/min, +119%) than the SHAM mouse (168 to 216 breaths/min = 48 breaths/min, +29%). The roll-off values at the end of the 5 min HXC were similar between the two mice (data not shown) as were the increases in Freq immediately upon return to room-air (RA at 15 sec, +131% and +114% for the SHAM and CSCX mouse respectively) and at 5 min (RA at 5 min, +131% and +100% for the SHAM and CSCX mouse respectively).

The changes in Freq, TV and MV in SHAM and CSCX mice in response to the 5 min hypoxic gas challenge and upon return to room-air are summarized in **Figure 15**. As seen in the top row of panels, exposure to HXC in SHAM mice elicited a typical initial increase in Freq that was subject to pronounced roll-off (left panel). The responses in CSCX mice were essentially similar compared to SHAM except that the rise in Freq was significantly higher at the 15 sec time-point (middle panel). The total increases in Freq in the three designated epochs (0-105 sec, 106-300 sec and 0-300 sec) were similar in the SHAM and CSCX mice (right panel). The return to room-air elicited the expected dramatic increase in Freq in SHAM mice and CSCX mice (left panel). The total responses were similar in both groups (**Table 5**). As seen in the middle row of panels, exposure to the HXC elicited immediate and sustained increases in TV that were similar in most aspects in the SHAM and CSCX mice except that the total increase recorded during the 106-300 sec epoch were higher in CSCX than SHAM mice (right panel). The return to room-air elicited the expected small initial increase in TV followed by gradual decline toward baseline in SHAM mice and CSCX mice (left panel). The total responses were similar in both groups (**Table 5**). As seen in the bottom row of panels, exposure to HXC in the SHAM mice elicited a typical initial increase in MV that was subject to a pronounced roll-off (left panel). The responses in CSCX mice were similar except that the rise in MV was significantly higher at the 15 sec time-point (middle panel). The total increases in MV in the three designated epochs were similar in the SHAM and CSCX mice. The return to room-air elicited the expected dramatic increase in MV in SHAM mice and CSCX mice (left panel). The total responses were similar in both groups (**Table 5**).

HYPOXIC CHALLENGES - INSPIRATORY TIME AND EXPIRATORY TIME

The changes in Ti and Te values in SHAM and CSCX mice in response to the 5 min HXC and upon return to room-air are summarized in **Figure 16**. Exposure to HXC in SHAM mice elicited initial decreases in Ti and Te that were subject to roll-off (left panels). The decreases in Ti and Te occurred more rapidly in CSCX mice (middle panels) but the overall (total) responses were similar in both groups (right panels). Return to room-air elicited a rapid transient decrease in Ti, but a rapid and sustained decrease in Te (left panels). The actual (left panels) and total responses (**Table 5**) were similar in SHAM and CSCX mice. As seen in **Table 5**, the total Te responses upon return to room-air fell into two groups, those in which total Te fell and those in which Te rose.

HYPOXIC CHALLENGES - INSPIRATORY TIME/EXPIRATORY TIME AND INSPIRATORY QUOTIENT

The changes in Ti/Te values and inspiratory quotients (Ti/(Ti+Te)) in SHAM and CSCX mice in response to the 5 min HXC and upon return to room-air are summarized in **Figure 17**. The resulting changes in Ti and Te during HXC (figure 3) translated into minor changes in Ti/Te values and inspiratory quotients in SHAM mice (left and middle panels) that were nevertheless significantly smaller in the CSCX mice (right panels). As seen in the left panels, the return to room-air resulted in substantial decreases in Ti/Te values and inspiratory quotients that gradually returned toward baseline values. As seen in **Table 5**, the changes in total Ti/Te and Ti/(Ti+Te) values upon return to room-air were similar in the SHAM and CSCX mice.

HYPOXIC CHALLENGES - END INSPIRATORY PAUSE, AND END EXPIRATORY PAUSE

The changes in EIP and EEP in SHAM and CSCX mice in response to the 5 min HXC and upon return to room-air are summarized in **Figure 18**. The hypoxic challenge elicited a prompt decrease in EIP in SHAM and CSCX mice (top left panel). The initial (top middle) and total responses (top right panel) were similar in SHAM and CSCX mice. The HXC elicited a fall in EEP of about 2 min in duration in SHAM and CSCX mice (bottom left panel). EEP returned to baseline values in the SHAM mice during the remainder of the hypoxic challenge but went above baseline in the CSCX mice. The initial decrease in EEP occurred faster in the CSCX mice (bottom middle panel) and the total changes in EEP (bottom right panel) reflected the biphasic changes described above. Return to room resulted in a gradual recovery of EIP toward baseline values, but substantial and variable increases in EEP (left panels). As seen in **Table 5**, the total EIP and EEP responses upon return to room-air were similar in the SHAM and CSCX mice.

HYPOXIC CHALLENGES - INSPIRATORY DRIVE AND EXPIRATORY DRIVE

The changes in inspiratory drive (TV/Ti) and expiratory drive (TV/Te) in the SHAM and CSCX mice in response to the 5 min HXC and upon return to room-air are summarized in **Figure 19**. The HXC elicited prompt and sustained increases in both inspiratory and expiratory drives (left panels) that occurred more rapidly in CSCX mice (middle panels) although the total responses were similar in both groups (right panels). Return to room-air elicited initial increase in both inspiratory and expiratory drives in SHAM mice and CSCX mice that gradually returned toward baseline (left panels). The total changes were similar in both groups (**Table 5**).

HYPOXIC CHALLENGES – PEAK INSPIRATORY AND EXPIRATORY FLOWS

The changes in PIF, PEF and PIF/PEF ratios in the SHAM and CSCX mice in response to the 5 min HXC and upon return to room-air are shown in **Figure 20**. The HXC elicited prompt and sustained increases in PIF and PEF in SHAM mice, with the PIF responses being of greater magnitude during the first half of the hypoxic challenge resulting in higher PIF/PEF ratios (left panels). The initial PIF responses during the HXC in the CSCX mice were similar to those in the SHAM mice, whereas the initial PEF responses were higher in CSCX mice, such that the expected increase in PIF/PEF ratio seen in SHAM mice did not occur in CSCX mice (middle panels). Total PIF responses to the HXC were similar in SHAM and CSCX mice, whereas the increases in PEF were higher in the CSCX mice compared to SHAM during each designated epoch (right panels). PIF/PEF ratios over the 106-300 sec and 0-300 sec epochs were markedly lower in CSCX mice than in SHAM mice (bottom right panel). The return to room-air elicited initial increase in PIF and PEF in SHAM mice and CSCX mice, and the changes resulted in sustained increases in PIF/PEF ratios (left panels). Total PIF, PEF and PIF/PEF responses were similar in both groups (**Table 5**).

HYPOXIC CHALLENGES – EF$_{50}$, R$_{PEF}$ AND RELAXATION TIME

The changes in EF$_{50}$, Rpef and RT values in the SHAM and CSCX mice in response to the 5 min hypoxic gas challenge and upon return to room-air are summarized in **Figure 21**. HXC elicited a robust increase in EF$_{50}$ that occurred more rapidly in the CSCX mice although the total responses were similar in the SHAM and CSCX mice (top row: left, middle and right panels). HXC-induced initial brief increases in Rpef in SHAM mice that were followed by sustained decreases (middle row: left, middle and right panels). The responses occurred more rapidly in the CSCX mice than SHAM mice, but the overall responses were similar in both groups. HXC elicited brief reductions in RT that occurred more rapidly in the CSCX mice than SHAM mice although the overall responses were similar in both groups (bottom row: left, middle and right panels). The return to room-air elicited prompt increases in EF$_{50}$ and Rpef that were associated with prompt decreases in RT (left panels). The overall changes in EF$_{50}$ were similar in both groups. The changes in RT in the SHAM and CSCX mice fell into two roughly equal numbers of mice, those in which RT was elevated and those in which RT was decreased. These changes were equivalent in the SHAM and CSCX mice. The changes in Rpef upon return to room-air also fell into two categories, roughly of equal numbers of mice, those in which Rpef was elevated and those in which RT was decreased. The values for those in which Rpef rose were much smaller in the CSCX mice than the SHAM mice (**Table 5**).

HYPOXIC CHALLENGES – REJECTION INDEX, REJECTION INDEX/FREQUENCY OF BREATHING, APNEIC PAUSES

The changes in RI, RI/Freq and Apneic Pause values in the SHAM and CSCX mice in response to the 5 min hypoxic gas challenge and upon return to room-air are summarized in **Figure 22**. The HXC elicited immediate, but short-lived increases in RI and RI/Freq values that were not accompanied by increases in the numbers of apneic pauses, which showed relatively transient decreases in the earlier stage of the HXC (left panels). The increases in RI and RI/Freq occurred more rapidly in the CSCX mice, but the overall changes were similar in the SHAM and CSCX mice (top and middle rows: middle and right panels). The changes in apneic pauses during the HXC were similar in the SHAM and CSCX mice (bottom row). The return to room-air was associated with rapid and substantial increases in RI, RI/Freq and apneic pauses (left panel) that were similar in magnitude in the SHAM and CSCX mice (**Table 5**).

DISCUSSION

Due to the multiplicity and often opposing effects of sympathetic nerves/catecholamines in the carotid bodies (see Introduction), the question as to how the absence of functional sympathetic input to the carotid bodies and other structures controlling ventilatory processes would affect baseline parameters and the responses to HXC was not easy to predict. The major finding that many of the ventilatory responses (e.g., increases Freq,

PEF and expiratory drive) during HXC occurred faster in CSCX mice suggests that activation of SCG sympathetic input to the carotid bodies of these mice normally blunts the initiation of these responses. Accordingly, it appears that loss of sympathetic input/catecholamine-induced suppression of carotid body activity (Kou et al., 1991; Pizarro et al., 1992; Prabhakar et al., 1993; Almaraz et al., 1997; Overholt and Prabhakar, 1999) out ways the loss of mechanisms which promote HXC responses including vasoconstriction in arterioles associated with primary glomus cells (Llados and Zapata, 1978; Eldridge and Gill-Kumar, 1980; Majcherczyk et al., 1980; Lahiri et al., 1981; Matsumoto et al., 1981; Gonsalves et al., 1984; Potter and McCloskey, 1987; Yokoyama et al., 2015). Deeper mechanistic insights would certainly come from studies designed to evaluate blood flow responses in the mouse carotid body and cerebral circulation in SHAM and CSCX mice during HXC and how systemic hemodynamic responses in these mice may influence the expression of the ventilatory responses

VENTILATORY RESPONSES DURING HYPOXIC GAS CHALLENGE

This study demonstrates that HXC in adult male C57BL6 mice elicits a complex array of ventilatory responses over and above what have been reported previously. In response to the HXC, the SHAM C57BL6 mice displayed an increase in Freq that was subject to roll-off and which was accompanied with decreases in Ti, Te, EEP, Rpef and RT that were also subject to roll-off. In contrast, the decrease in EIP was sustained throughout HXC. HXC was also associated with robust and sustained increases in TV, MV, PIF, PEF, EF_{50} and inspiratory and expiratory drives. The question arises as to why some ventilatory parameters in C57BL6 mice are subject to roll-off whereas others are not. First, it should be noted that the C57BL6 mouse is widely considered to be a "normal" healthy model to study the physiology of cardiorespiratory systems, and is used to generate genetically-engineered mice to study the mechanisms involved in cardiorespiratory-thermoregulatory control processes, including responses to hypoxic gas challenges (Tankersley et al., 2002; Campen et al., 2004, 2005; Tewari et al., 2013; Palmer et al., 2013a,b; Gaston et al., 2014). Despite considerable normal physiology, the C57BL6 mouse is of major interest to sleep-apnea researchers because it displays disordered breathing (irregular breathing patterns, including apneas and sighs) and cardiovascular disturbances during sleep and wakefulness, and shows disordered breathing upon return to room-air after exposure to HXC (Han et al., 2000, 2001, 2002; Tagaito et al., 2001; Yamauchi et al., 2008a,b,c, 2010; Getsy et al., 2014). The genetic (Tankersley et al., 1994, 2000, 2002; Tankersley, 2001, 2003; Han et al., 2001, 2002; Tagaito et al., 2001; Yamauchi et al., 2008b) and neurochemical processes (Tankersley et al., 2002; Price et al., 2003; Groeben et al., 2005; Yamauchi et al., 2008a,c, 2010; Moore et al., 2012, 2014), underlying the breathing patterns of C57BL6 mice, and th responses to HXC are well studied. The potential role of structural differences in the carotid bodies has also been studied (Yamaguchi et al., 2003, 2006; Chai et al., 2011). The breathing patterns of C57BL6 mice, and their responses to HXC have a strong genetic component but there is at present no explanation for why some ventilatory components, such as ventilatory timing (e.g., Freq, EEP) and mechanics (e.g., Rpef, RT) are subject to roll-off whereas other timing (e.g., EIP) and mechanics (e.g., PIF, PEF, EF_{50}) are not. Regardless of the explanation, understanding the importance of each of these ventilatory responses to HXC will help us better understand the processes by which breathing disorders occur in disease states and point to therapeutic strategies.

Resting ventilatory parameters (19 directly recorded or calculated variables) were similar in the SHAM and CSCX mice. This would suggest that (presumed) loss of input from the SCG to structures controlling breathing, such as the carotid bodies, upper airway and brainstem structures (see Introduction), do not obviously effect ventilatory timing or mechanics or the quality of breathing (e.g., occurrence of non-eupneic breathing events, such as apneic pauses) in C57BL6 mice. Again, the caveat is, that these studies were performed only 4 days after CSCX, and it would seem possible that changes in baseline ventilatory performance would occur at longer post-CSCX time-points. Nonetheless, there were numerous important differences in the responses of CSCX and SHAM mice to the HXC. For example, the increases in Freq (and associated decreases in Ti, Te, EEP and RT) and the increases in MV, inspiratory and expiratory drives, PEF and EF_{50} occurred more quickly in CSCX mice than in SHAM mice. Although both PIF and PEF increased during HXC, the PIF/PEF ratio fell because the rise in PEF exceeded that of the rise in PIF. The PIF/PEF ratio fell more dramatically, and to a greater extent, in the CSCX mice because of the exaggerated rise in PEF in the CSCX mice. The overall (total) responses in the SHAM and CSCX mice to the HXC were similar to one another with some important exceptions. Specifically, the overall increases in tidal volume during the latter half of the hypoxic challenge were greater in the CSCX mice, and as mentioned, the PIF/PEF ratio fell and to a greater extent in the CSCX mice. Moreover, the total decreases in Ti/Te and respiratory quotient (Ti/(Ti+Te) observed in the SHAM mice were absent in CSCX mice. These findings clearly demonstrate that CSC-SCG input to respiratory control structures influence the ventilatory responses to HXC in C57BL6 mice.

VENTILATORY RESPONSES UPON RETURN TO ROOM-AIR

Following exposure to HXC, the return to room-air resulted in respiratory patterns that can be classified as short-term potentiation, in which ventilation remains elevated (Powell et al., 1998; Getsy et al., 2014) or post-hypoxic frequency decline, in which breathing frequency falls below baseline (Dick and Coles, 2000). Our C57BL6 mice displayed robust short-term potentiation upon return to room air, accompanied by a substantial prolonged phase of disordered breathing (e.g., elevated rejection index). The mechanisms responsible for post-HXC disordered breathing have received considerable investigation, and at present, evidence is in favor of disturbances in central signaling (Wilkinson et al., 1997; Strohl, 2003) including, the pons area of the brainstem (Coles and Dick, 1996; Dick and Coles, 2000) rather than processes within the carotid bodies (Vizek et al., 1987; Brown et al., 1993), even though it is evident that carotid body chemoafferents play a vital role in the expression of sleep apnea (Smith et al., 2003). The post-HXC responses were similar in our SHAM and CSCX mice suggesting that diminished input to the SCG and (presumably decreased activity of SCG cells) do not have an obvious impact on the post-HXC responses, including the disordered breathing. The one exception was that total Rpef responses (positive rather than negative responders) (**Table 2**) after return to room-air were smaller in the CSCX mice than in the SHAM mice. This suggests that CSC-SCG activity is normally a positive factor in achieving maximal Rpef upon recovery from a HXC challenge in C57BL6 mice. The major differences with respect to the changes in Freq, MV, EEP, rejection index

(Rinx), Rinx/Freq did occur during the first 15 sec although other differences between the groups occurred at 30 sec for relaxation time and inspiratory drive; 45 sec for Rpef, 15 and 30 sec for Inspiratory time and expiratory time; 30, 45 and 60 sec for PEF, PEF/PIF and EF_{50}; and 15, 30, 45 and 60 sec for expiratory drive.

CONCLUSION

Our data show that under baseline (normoxic) environmental conditions, the potential loss of active sympathetic input to the carotid bodies (via CSCX-induced quiescence of postganglionic projections to the carotid body) does not have a noticeable effect on ventilatory parameters, which suggests that resting activity (e.g., neurotransmitter release) of carotid body glomus cells is not altered in a way that would lead to activation of carotid body chemoafferents and therefore enhancement of breathing. In contrast, our data show that CSCX does alter the HXC ventilatory response and therefore suggests that diminished SCG input to the carotid bodies (or other targets such as those in the brainstem) does influence the ability of glomus cells and/or other neuronal structures to respond to HXC. Planned studies involving bilateral transection of ganglioglomerular nerves (that project only to the carotid bodies and carotid sinus) will help to establish the relevant neuronal pathways/mechanisms. Overall, this novel data suggest that the CSC may normally provide inhibitory input to peripheral (e.g., carotid bodies) and central (e.g., brainstem) structures that are involved in the ventilatory responses to hypoxic gas challenge in C57BL6 mice. Moreover, the results of our CSCX studies lend support to the concept that a loss of CSC-SCG activity may be involved in the etiology of ventilatory disorders, such as sleep-disordered breathing. With respect to mechanistic insights provided by our data, it would be reasonable to assume that postganglionic sympathetic nerves innervating the ipsilateral carotid bodies (and other targets such as those within the brain) would be quiescent following CSCX as a result of the loss of preganglionic input to the SCG. Whether the presumed diminution of sympathetic activity alters the expression of functional proteins (i.e., tyrosine hydroxylase, catecholamine-containing vesicles, fusion proteins mediating vesicular exocytosis) within the sympathetic nerve terminals themselves and/or the target tissues such as type 1 glomus cells in the carotid bodies needs to be addressed. This is especially important since these post-transection adaptations in protein expression, while not being noticeable at rest (i.e., under normoxia) may have a direct effect on the ability of glomus cells (for example) to respond to the hypoxic challenge and/or secrete neurotransmitters. Our companion manuscript (Getsy et al., 2020) demonstrates that removal of the SCG has dramatically augmented effects on HXC compared to CSCX. This raises important questions as to whether HXC may directly alter the activity of SCG neurons independently of the CSC input (Getsy et al., 2020). One key question pertains to which of the SCG projections normally driven by the CSC are responsible for mediating the neuromodulatory effects of the CSC on the processes that drive ventilatory responses to HXC. As detailed in the Introduction section, the direct projections of the SCG to structures that control ventilation are extensive and include projections to the carotid bodies via the GGN (branch of the external carotid nerve). In order to better determine which of the postganglionic projections of the SCG regulate the ventilatory responses to HXC, we are currently planning to perform studies in mice in which the major postganglionic branches of the SCG, namely the internal carotid nerves, external carotid nerves and GGN are transected. We intend to perform these studies 4, 14 and 30 days post-transection in C57BL6 mice and in other strains, such as Swiss-Webster and A/J mice (Getsy et al., 2014) to determine temporal and genetic aspects of the role of the CSC-SCG complex in the control of ventilation and the responses to HXC.

The data demonstrate that the primary effect of CSCX appears to be changes in the immediate responsiveness to the HXC. For example, the increases in Freq (and associated decreases in Ti, Te and EEP), MV, expiratory drive and rejection index occurred more rapidly in CSCX mice than SHAM mice (changes at 15 sec were significant in CSCX mice but not SHAM mice and between-group differences for expiratory drive were maintained at 15, 30, 45 and 60 sec). In contrast, between-group differences in the responses of relaxation time, Rpef, PIF, PEF, PEF/PIF, EF_{50}, and inspiratory drive, were evident at 30, 45 or 60 sec with between-group differences for PEF, PEF/PIF and EF_{50} being evident for 30, 45 and 60 sec. Important to the interpretation of the effects of CSCX, it was evident that the initial increases in TV and PIF during HXC were similar in SHAM and CSCX mice. Taken together, it is apparent that the loss of postganglionic SCG input to structures such as the carotid body has a strong impact on ventilatory performance in C57B6 mice. The finding that the decreases in Ti and Te were larger in the CSCX mice than the SHAM mice suggests that SCG input to neural structures regulating ventilatory timing events equally affect inspiratory and expiratory control processes. However, with respect to flow parameters it was evident that (a) the increases in TV were similar in SHAM and CSCX mice, (b) the responses of PIF and inspiratory drive in CSCX mice were minimally different from SHAM mice whereas the changes in PEF, EF_{50}, PEF/PIF and expiratory drive were substantially different between the groups. As such, CSCX appears to have much more of an influence on peak expiratory flows than inspiratory flows. Whether this pattern of effects is due to altered carotid body function must await more definitive studies in which, for example, the effects of bilateral ganglioglomerular nerve transection are investigated

The data in the present manuscript and in that of its companion which investigated the effects of SCGX (Getsy et al., 2020b) provide the beginnings of understanding how the loss of preganglionic and/or postganglionic fibers in the CSC-SCG complex affect resting ventilatory parameters and the responses to HXC. On-going studies will extend our investigations by determining how expression of proteins in sympathetic nerve terminals (e.g., tyrosine hydroxylase, fusion proteins, vesicular stores of norepinephrine) and primary glomus cells (tyrosine hydroxylase, voltage-gated Na^+, K^+ and Ca^{2+}-channels) change after CSCX, and also by performing ventilatory studies in mice with (a) transection of the left and right internal carotid nerves (major postganglionic SCG trunks), (b) transection of the left and right external carotid nerves (the other major postganglionic SCG trunks) and (c) transection of the left and right ganglioglomerular nerves (a branch of the external carotid nerve), which project only to the carotid bodies and carotid sinus.

Table 4

Baseline parameters in sham-operated (SHAM) mice and in mice with bilateral transection of the cervical sympathetic chain (CSCX)

Parameter	SHAM	CSCX
Number of mice	11	12
Age, days	101 ± 1	102 ± 1
Body Weight, grams	27.2 ± 0.5	27.5 ± 0.8
Frequency (breaths/min)	196 ± 5	188 ± 6
Tidal Volume (TV, ml)	0.149 ± 0.007	0.157 ± 0.014
Minute Ventilation (ml/min)	28.9 ± 1.1	28.6 ± 1.8
Inspiratory Time (Ti, sec)	0.112 ± 0.003	0.114 ± 0.003
Expiratory Time (Te, sec)	0.207 ± 0.006	0.220 ± 0.009
End Inspiratory Pause (EIP, msec)	2.90 ± 0.08	2.85 ± 0.07
End Expiratory Pause (EEP, msec)	34.5 ± 3.8	32.7 ± 6.1
Ti/Te	0.551 ± 0.016	0.526 ± 0.015
Ti/(Ti + Te)	0.353 ± 0.007	0.343 ± 0.006
Peak Inspiratory Flow (PIF, ml/sec)	2.38 ± 0.09	2.44 ± 0.16
Peak Expiratory Flow (PEF, ml/sec)	1.51 ± 0.06	1.42 ± 0.09
PIF/PEF	1.59 ± 0.05	1.74 ± 0.07
Expiratory flow at 50% exhaled TV (EF_{50}, ml/sec)	0.072 ± 0.004	0.071 ± 0.003
Relaxation Time (RT, sec)	0.100 ± 0.003	0.103 ± 0.004
Rate of achieving PEF (Rpef)	0.138 ± 0.13	0.166 ± 0.019
Inspiratory Drive (TV/Ti, ml/sec)	1.36 ± 0.06	1.38 ± 0.09
Expiratory Drive (TV/Te, ml/sec)	0.72 ± 0.03	0.70 ± 0.04
Rejection Index (RI, %)	15.4 ± 1.9	15.2 ± 1.9
(Rejection Index/Frequency) x 100	7.9 ± 0.9	8.3 ± 1.2
Apneic Pauses (Te/RT)-1	1.07 ± 0.03	1.15 ± 0.05

The data are presented as mean $\pm$ SEM. There were no between-group differences for any parameter (P > 0.05, for all comparisons).

Table 5

Total changes that occurred during the first 15 min upon return to room-air

Parameter	SHAM	CSCX
Number of mice	11	12
Frequency (%)	+33.0 ± 11.2	+39.4 ± 13.2
Tidal Volume (TV, %)	+33.3 ± 7.1	+39.8 ± 7.4
Minute Ventilation (%)	+87.1 ± 22.1	+105.3 ± 25.8
Inspiratory Time (Ti, %)	-26.5 ± 5.5	-28.3 ± 5.0
Expiratory Time (Te, %)	+3.2 ± 7.9	+1.2 ± 8.2
+Te,%	+17.2 ± 3.4 (6)	+18.4 ± 3.1 (8)
-Te, %	-26.6 ± 4.0 (5)	-33.3 ± 5.1 (4)
Ti/Te, %	-25.3 ± 2.3	-25.2 ± 3.5
Ti/(Ti + Te), %	-18.9 ± 1.6	-19.0 ± 2.7
End Inspiratory Pause (%)	-12.9 ± 2.0	-10.7 ± 2.0
End Expiratory Pause (EEP, msec)	+181 ± 29	+298 ± 61
Peak Inspiratory Flow (PIF, %)	+106 ± 19	+122 ± 21
Peak Expiratory Flow (PEF, %)	+92 ± 20	+128 ± 28
PIF/PEF, %	+13.5 ± 3.4	+6.9 ± 4.8
Expiratory flow at 50% exhaled TV (EF$_{50}$, %)	+89 ± 22	+114 ± 30
Relaxation Time (RT, %)	-0.3 ± 6.4	-4.1 ± 7.3
+ RT,%	+16.6 ± 3.0 (5)	+24.6 ± 7.9 (7)
- RT, %	-14.3 ± 3.7 (6)	-16.6± 3.0 (5)
Rate of achieving PEF (Rpef, %)	+27.1 ± 13.1	+21.9 ± 3.6
+ Rpef,%	+61 ± 7 (6)	+21.9 ± 3.6* (5)
- Rpef, %	-14 ± 5 (5)	-23.6 ± 2.5 (7)
Inspiratory Drive (TV/Ti, %)	+114 ± 21	+133 ± 24
Expiratory Drive (TV/Te, %)	+59 ± 19	+78 ± 25
Rejection Index (RI, %)	+227 ± 50	+200 ± 33
(Rejection Index/Frequency) x 100 (%)	+123 ± 27	+132 ± 28
Apneic Pauses (Te/RT)-1 (%)	+8.2 ± 7.0	+13.8 ± 5.4
+ Te/RT)-1,%	+26.7 ± 6.5 (5)	+19.8 ± 3.3 (10)
- Te/RT)-1, %	-7.2 ± 2.7 (6)	-16.5 ± 5.8 (2)

The data are presented as mean ± SEM. †P < 0.05, CSCX versus SHAM. The values in parentheses equal the number of mice in the sub-groups for Te, RT, Rpef and (Te/RT)-1.

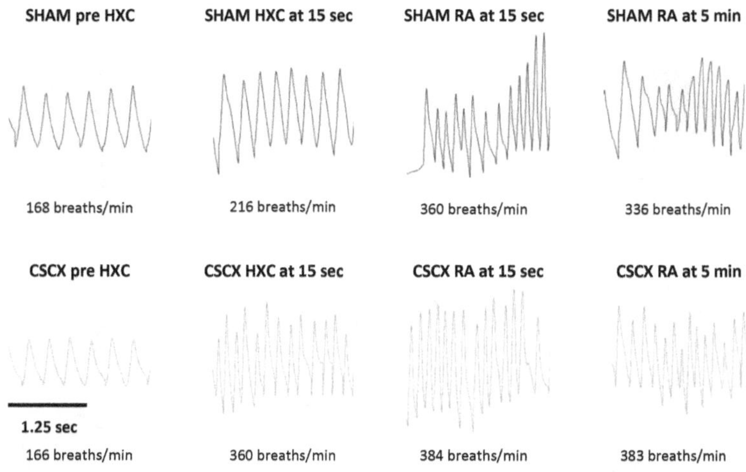

Figure 14. Examples of respiratory waveforms during various stages of the experiment in a sham-operated (SHAM) mouse and in a mouse with bilateral cervical sympathetic chain transection (CSCX). A time bar (1.25 sec) is shown in the bottom left of the figure. HXC, hypoxic gas challenge. RA, room-air.

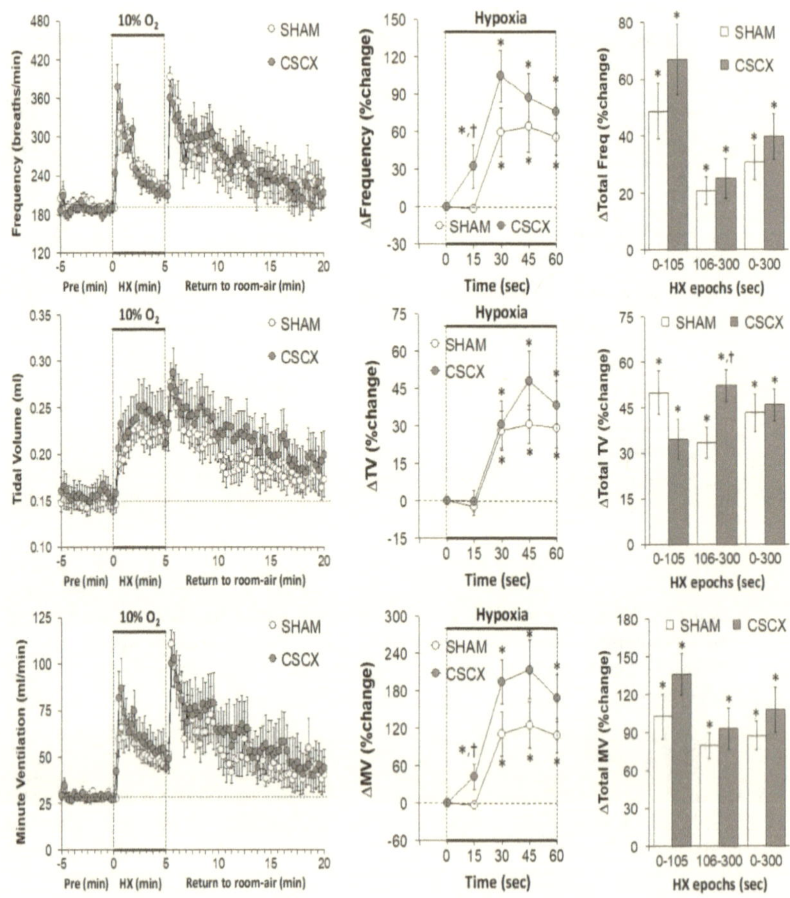

Figure 15. Left panels: Frequency of breathing, tidal volume and minute ventilation values before, during and after a 5 min hypoxic (HX, 10% O_2, 90% N_2) gas challenge in sham-operated (SHAM) mice and in mice with bilateral transection of the cervical sympathetic chain (CSCX). **Middle Panels:** Responses (expressed as % of pre-values) during the first minute of HX challenge in SHAM and CSCX mice. **Right panels:** Total responses (sum of all % changes from pre) during the first 105 seconds, between 106-300 sec and between 0-300 sec. Data are expressed as mean ± SEM. *P < 0.05, significant response. †P < 0.05, CSCX versus SHAM. There were 11 mice in the SHAM group and 12 mice in the CSCX group.

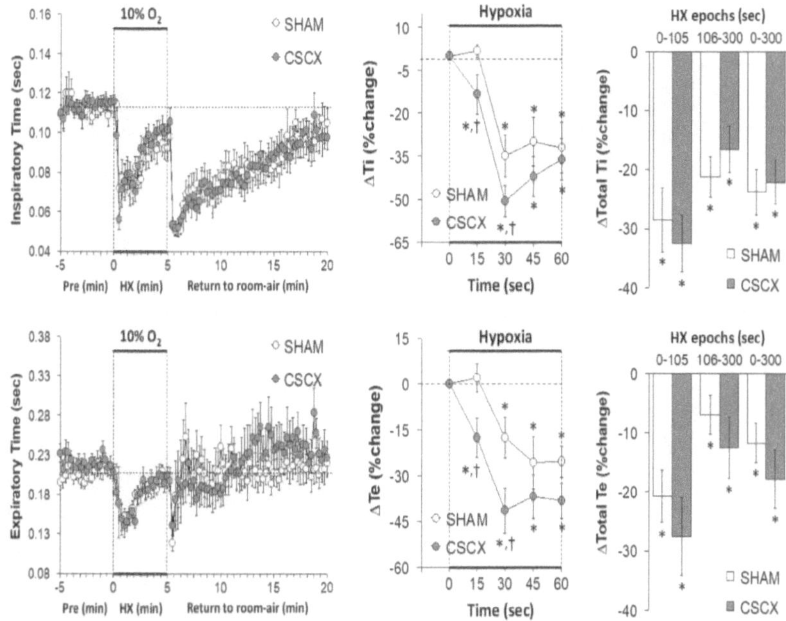

Figure 16. **Left panels:** Inspiratory time (Ti) and expiratory time (Te) values before, during and after a 5 min hypoxic (HX, 10% O_2, 90% N_2) gas challenge in sham-operated (SHAM) mice and in mice with bilateral transection of the cervical sympathetic chain (CSCX). **Middle Panels:** Responses (expressed as % of pre-values) during the first minute of HX challenge in SHAM and CSCX mice. **Right panels:** Total responses (sum of all % changes from pre) during the first 105 seconds, between 106-300 sec and between 0-300 sec. Data are expressed as mean ± SEM. *P < 0.05, significant response. †P < 0.05, CSCX versus SHAM. There were 11 mice in the SHAM group and 12 mice in the CSCX group.

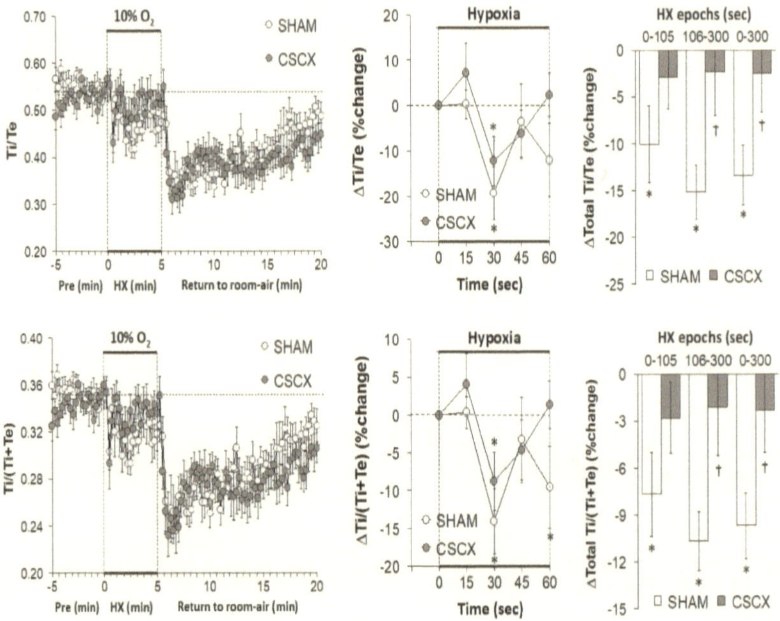

Figure 17. Left panels: Inspiratory time/expiratory time (Ti/Te) and inspiratory time/(inspiratory time + expiratory time) ratios [(Ti/(Ti + Te)] before, during and after a 5 min hypoxic (HX, 10% O_2, 90% N_2) gas challenge in sham-operated (SHAM) mice and in mice with bilateral transection of the cervical sympathetic chain (CSCX). **Middle Panels:** Responses (expressed as % of pre-values) during the first minute of HX challenge in SHAM and CSCX mice. **Right panels:** Total responses (sum of all % changes from pre) during the first 105 seconds, between 106-300 sec and between 0-300 sec. Data are expressed as mean ± SEM. *$P < 0.05$, significant response. †$P < 0.05$, CSCX versus SHAM. There were 11 mice in the SHAM group and 12 mice in the CSCX group.

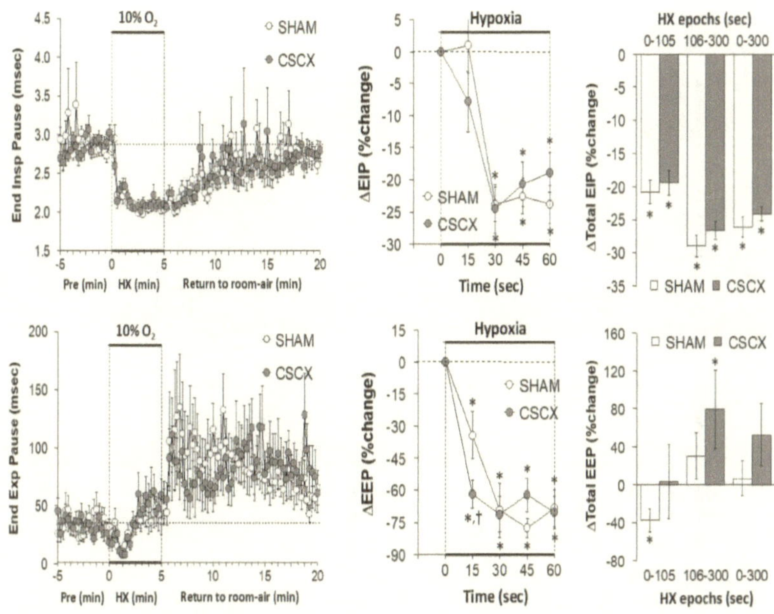

Figure 18. **Left panels:** End inspiratory pause (EIP) and end expiratory pause (EEP) values before, during and after a 5 min hypoxic (HX, 10% O_2, 90% N_2) gas challenge in sham-operated (SHAM) mice and in mice with bilateral transection of the cervical sympathetic chain (CSCX). **Middle Panels:** Responses (expressed as % of pre-values) during the first minute of HX challenge in SHAM and CSCX mice. **Right panels:** Total responses (sum of all % changes from pre) during the first 105 seconds, between 106-300 sec and between 0-300 sec. Data are expressed as mean ± SEM. *$P < 0.05$, significant response. †$P < 0.05$, CSCX versus SHAM. There were 11 mice in the SHAM group and 12 mice in the CSCX group.

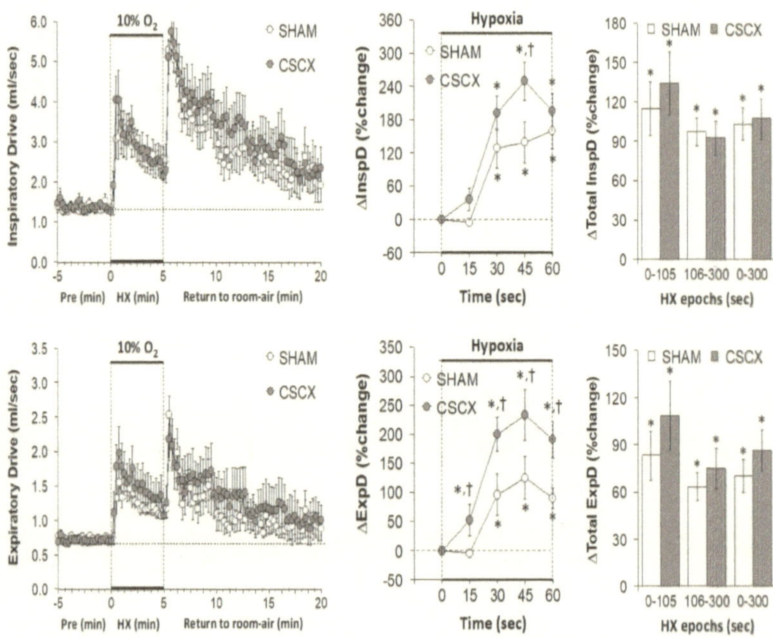

Figure 19. Left panels: Inspiratory Drive (TV/Ti) and Expiratory Drive (TV/Te) before, during and after a 5 min hypoxic (HX, 10% O_2, 90% N_2) gas challenge in sham-operated (SHAM) mice and in mice with bilateral transection of the cervical sympathetic chain (CSCX). **Middle Panels:** Responses (expressed as % of pre-values) during the first minute of HX challenge in SHAM and CSCX mice. **Right panels:** Total responses (sum of all % changes from pre) during the first 105 seconds, between 106-300 sec and between 0-300 sec. Data are expressed as mean ± SEM. *P < 0.05, significant response. †P < 0.05, CSCX versus SHAM. There were 11 mice in the SHAM group and 12 mice in the CSCX group.

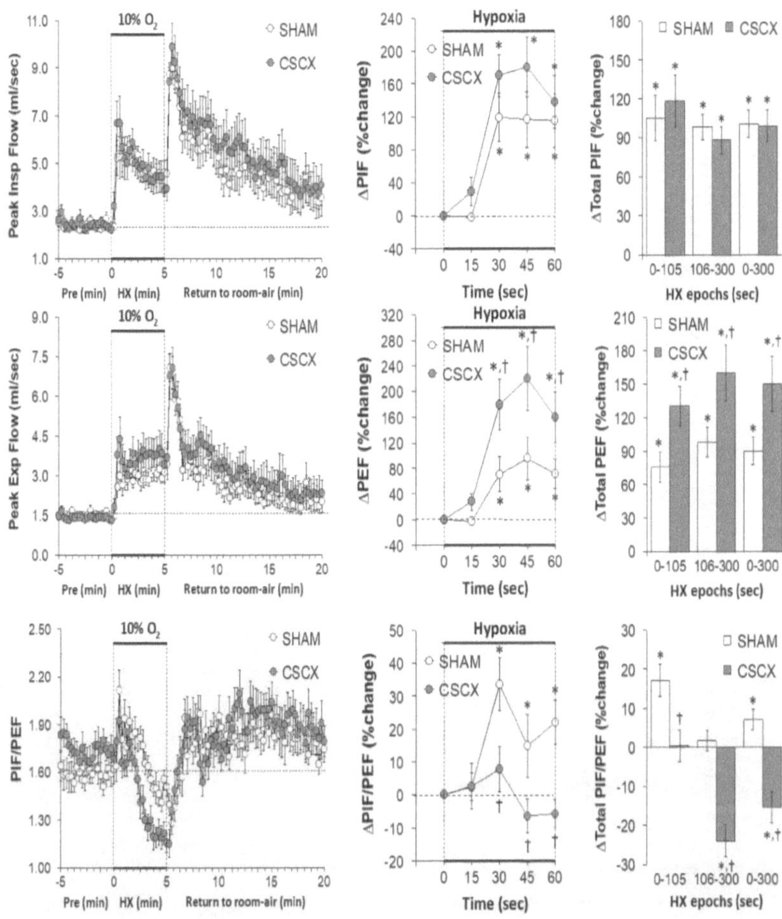

Figure 20. Left panels: Peak Inspiratory flow (PIF) and Peak Expiratory Flow (PEF) values and PIF/PEF ratios before, during and after a 5 min hypoxic (HX, 10% O_2, 90% N_2) gas challenge in sham-operated (SHAM) mice and in mice with bilateral transection of the cervical sympathetic chain (CSCX). **Middle Panels:** Responses (expressed as % of pre-values) during the first minute of HX challenge in SHAM and CSCX mice. **Right panels:** Total responses (sum of all % changes from pre) during the first 105 seconds, between 106-300 sec and between 0-300 sec. Data are expressed as mean ± SEM. *P < 0.05, significant response. †P < 0.05, CSCX versus SHAM. There were 11 mice in the SHAM group and 12 mice in the CSCX group.

99

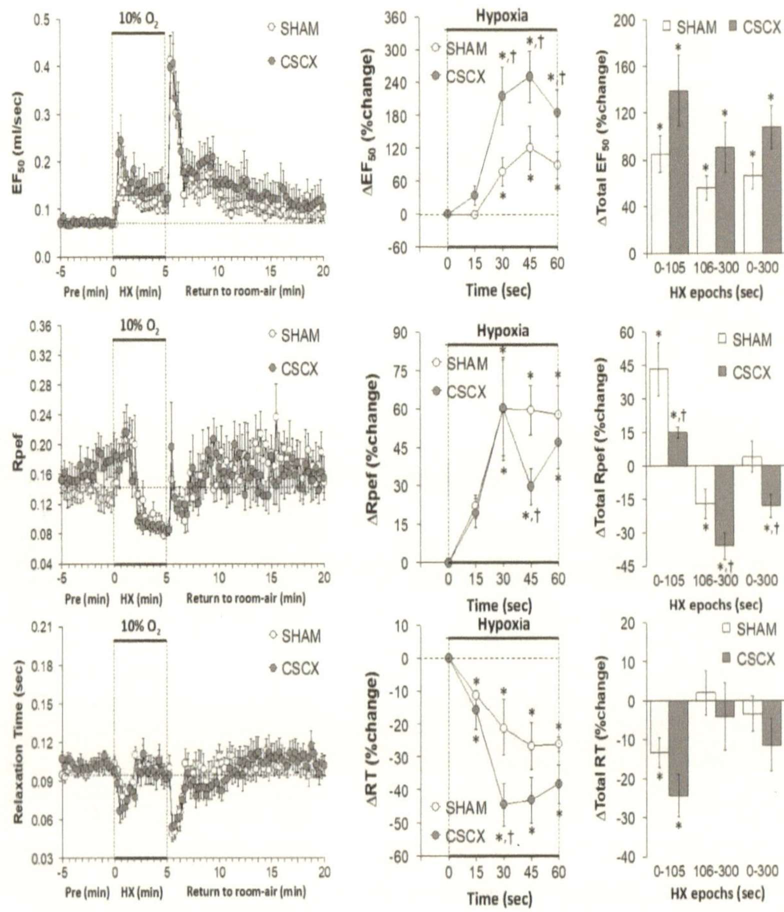

Figure 21. Left panels: EF_{50}, Rpef and Relaxation Time values before, during and after a 5 min hypoxic (HX, 10% O_2, 90% N_2) gas challenge in sham-operated (SHAM) mice and in mice with bilateral transection of the cervical sympathetic chain (CSCX). **Middle Panels:** Responses (expressed as % of pre-values) during the first minute of HX challenge in SHAM and CSCX mice. **Right panels:** Total responses (sum of all % changes from pre) during the first 105 seconds, between 106-300 sec and between 0-300 sec. Data are expressed as mean ± SEM. *P < 0.05, significant response. †P < 0.05, CSCX versus SHAM. There were 11 mice in the SHAM group and 12 mice in the CSCX group.

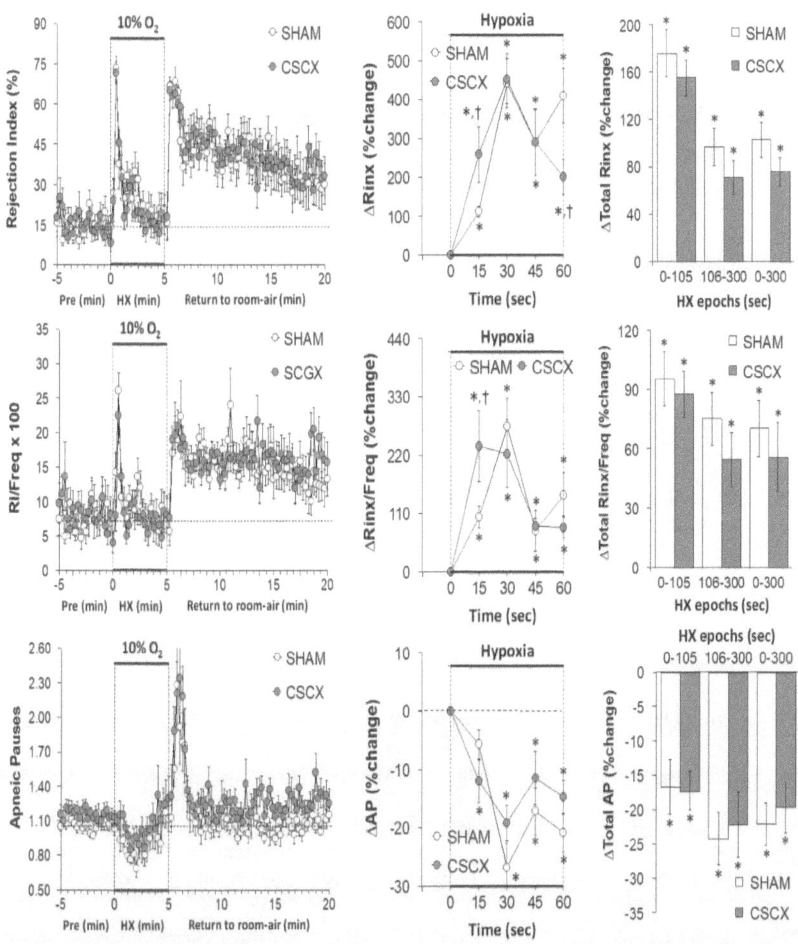

Figure 22. Left panels: Rejection Index (RI, Rinx) values, RI/frequency ratios and Apneic Pauses (AP) before, during and after a 5 min hypoxic (HX, 10% O_2, 90% N_2) gas challenge in sham-operated (SHAM) mice and in mice with bilateral transection of the cervical sympathetic chain (CSCX). **Middle Panels:** Responses (expressed as % of pre-values) during the first minute of HX challenge in SHAM and CSCX mice. **Right panels:** Total responses (sum of all % changes from pre) during the first 105 seconds, between 106-300 sec and between 0-300 sec. Data are expressed as mean ± SEM. *P < 0.05, significant response. †P < 0.05, CSCX versus SHAM. There were 11 mice in the SHAM group and 12 mice in the CSCX group.

101

CHAPTER IV: THE SUPERIOR CERVICAL GANGLIA MODULATE VENTILATORY RESPONSES TO HYPOXIA INDEPENDENTLY OF PREGANGLIONIC DRIVE FROM THE CERVICAL SYMPATHETIC CHAIN

INTRODUCTION

The cell bodies of postganglionic sympathetic neurons (Llewellyn-Smith et al., 1998; Tang et al., 1995a,b,c; Rando et al., 1981) and small intensely fluorescent (SIF) cells (Takaki et al., 2015; Zaidi and Matthews, 2013; McDonald, 1983a,b) within the left and right superior cervical ganglia (SCG) receive their preganglionic input from thoracic spinal cord (T1-T4) nerves that course within the ipsilateral cervical sympathetic chains (CSC). The majority of postganglionic nerves project from the SCG via two main trunks, the external (ECN) and internal (ICN) carotid nerves (Savastano et al., 2010; Asomoto, 2004; Buller and Bolter, 1997; Bowers and Zigmond, 1979). In addition, postganglionic SCG fibers in the ECN branch into the ganglioglomerular nerve (GGN) to innervate (a) numerous structures in the carotid bodies including the vasculature, chemoreceptor afferent nerve terminals and type 1 glomus cells (Savastano et al., 2010; Asamoto, 2004; Ichikawa, 2002 Torrealba and

Claps, 1988; Verna et al., 1984; McDonald, 1983a,b; McDonald and Mitchell, 1981; Bowers and Zigmond, 1979; Zapata et al., 1969; Biscoe and Purves, 1967), and (b) to the terminals of baroreceptor afferent nerves within the carotid sinus (Buller and Bolter, 1993; Felder et al., 1983; Bolter and Ledsome, 1976; Rees, 1967; Floyd and Neil, 1952). The above SCG projections in addition to those directed to nuclei within the brainstem, such the nucleus tractus solitarius (NTS) and hypothalamus (Mathew, 2007; Hughes-Davis et al., 2005; Esquifino et al., 2004; Westerhaus and Loewy, 1999; Wiberg and Widenfalk, 1993; Saavedra, 1985; Gallardo et al., 1984; Cardinali et al., 1982, 1981a,b) as well as the tongue and upper airway (Oh et al., 2006; Wang and Chiou, 2004; Hisa et al., 1999; O'Halloran et al., 1996, 1998; Kummer et al., 1992; Flett and Bell, 1991), support the possibility that the SCG is vital to the integration of cardiorespiratory function.

Until recently, no *in vivo* studies have characterized the effects of transection of the CSC or surgical removal of the SCG on breathing or the ventilatory responses that occur during hypoxic gas exposures. Our companion manuscript (Getsy et al., 2021) reported that the responses of several breathing parameters that occurred during a 5-min challenge with a hypoxic gas mixture (HXC, 10% O_2, 90% N_2) were substantially modified in C57BL6 mice that had undergone bilateral CSC transection (CSCX) 4 days previously. First, breathing parameters at rest in the CSCX mice were similar to those of mice that underwent sham-operation (SHAM). However, we found that the HXC-induced increases in frequency of breathing (Freq) and accompanying decreases in inspiratory time (Ti), expiratory time (Te), end expiratory pause (EEP, pause between end of expiration and start of inspiration) and relaxation time (RT, decay of respiration to 36% of maximum peak inspiratory flow), as well as increases in minute ventilation (MV), expiratory drive, and expiratory flow at 50% exhaled tidal volume (EF_{50}) occurred faster (i.e., reached peak values more quickly) in CSCX mice than in SHAM mice, although the total responses that occurred during the HXC were similar in CSCX and SHAM mice. In addition, the total increase in tidal volume (TV) during the latter half of the HXC was higher in CSCX mice than SHAM mice whereas the early and total increases in peak inspiratory flow (PIF) were higher in the CSCX mice. These findings suggest that in C57BL6 mice, the CSC input to the SCG drives SCG cells to provide inhibitory control to central structures (e.g., nuclei in the brainstem) and/or the carotid bodies that mediate the ventilatory responses to HXC. The reasons for choosing C57BL6 mice for this and our companion study (Getsy et al., 2021), including descriptions of the genetic, morphological, neurophysiological and neurochemical factors that participate in the expression of the ventilatory responses of C57BL6 mice to HXC are described in detail by Getsy et al., (2021).

As mentioned, the SCG contains postganglionic sympathetic cell bodies that receive their preganglionic (cholinergic) input from nerve fibers in the CSC (Savastano et al., 2010; Asomoto, 2004; Buller and Bolter, 1997; Bowers and Zigmond, 1979), and SIF cells that also receive preganglionic input via the CSC (Takaki et al., 2015; Zaidi and Matthews, 2013; McDonald, 1983a,b). The evidence that neural activity in the CSC and GGN increases during HXC (Matsumoto et al., 1987, 1986; Lahiri et al., 1986), suggests that HXC elicits preganglionic nerve-induced activation of postganglionic SCG neurons and SIF cells. As such, a carotid sinus (chemoreceptor afferent) nerve-brainstem-descending spinal cord pathway may be activated during HXC, which in turn increases the activity of preganglionic CSC nerves projecting to SCG cells that in turn provide input to key ventilatory structures, such as the carotid body via the GGN branch. Thus, it can be hypothesized that mice with bilateral superior cervical ganglionectomy (SCGX) would present with similar responses to HXC as those with bilateral CSCX. It is important to note that while it appears that principal SCG cells are not hypoxia-sensitive (Bernardini et al., 2020; Gao et al., 2019; Buckler and Turner, 2013; Nunes et al., 2010, 2012; Rigual et al., 1999) there is convincing evidence that SIF cells/interneurons in the SCG directly respond to hypoxic challenges (Strosznajder et al., 1997; Dinger et al., 1993; Brokaw and Hansen, 1987; Hanson et al., 1986). Indeed, Dinger et al (1993) clearly demonstrated that SIF cells in the SCG are selectively activated by hypoxic stimuli and that the neurochemical responses of these cells are comparable to those observed with chemosensory type I (glomus) cells of the carotid body.

The objectives of the present study were to use whole body plethysmography (Palmer et al., 2015; Gaston et al., 2014; Getsy et al., 2014; Palmer et al., 2013a,b;) in order to (1) compare resting ventilation in adult male C57CL6 mice in which both SCG were removed 4 days previously to those that were sham-operated (SHAM), and (2) compare their ventilatory responses to HXC. The present study demonstrates that SCGX mice displayed numerous altered responses to HXC that also occurred in CSCX mice, but also displayed numerous changes in HXC responses that CSCX mice did not, including diminished total increases in minute ventilation. As such, the major finding of this study is that bilateral removal of the SCG does not elicit the same effects as CSC transection

(Getsy et al., 2021), which suggests that hypoxia exerts multiple changes in neural signaling in the CSC-SCG pathway. These data raise the possibility that HXC directly activates sub-populations of SCG cells, including SIF cells and/or sub-populations of postganglionic neurons independently of increased input from the CSC, and that this SCG input to cardiorespiratory structures is of vital importance to hypoxic ventilatory responses.

RESULTS

RESTING PARAMETERS

The ages and body weights of the mice and their baseline (resting) ventilatory parameters prior to exposure to HXC are summarized in **Table 6**. The SHAM and CSCX groups each contained 14 mice. There were no differences in the ages or body weights of the two groups of mice (P > 0.05, for each comparison) and therefore it was not necessary to correct for body weights for the volume parameters (i.e., TV, PIF and PEF). No between-group differences were found for any variable (P > 0.05, for all comparisons) except for the greater degree of apneic pauses occurring in SCGX mice (P < 0.05). The left panel of each row in the figures to be described below will display the values recorded before, during the HXC (10% O_2, 90% N_2 for 5 min) and foowing reintroduction of room-air. The arithmetic changes recorded over the first 60 sec of HXC will be displayed in the middle panels of each row. The total response (% change from pre-values) during 3 epochs of the 300 sec HXC, 0-105 sec, 106-300 sec and 0-300 sec, will be displayed in the right panels of each row. These epochs best represented the clearest differences between the SCGX and SHAM mice.

Prior to introduction of the HXC, the mice were calm and still although they would occasionally groom or move about the chamber. Upon introduction of the HXC, the mice would not change behavior (i.e., there was no obvious direct response of the mice when switching to the HX gas) and became progressively more still as the HXC progressed, as described previously (Palmer et al., 2015; Getsy et al., 2014; Gaston et al., 2014; Palmer et al., 2013a,b). Moreover, the mice returned to normal levels of behaviors such as grooming upon reintroduction of room-air and accordingly, the long-lasting increase in RI upon return to room-air was not simply explained by an increase in behaviors such as locomotion, rearing, grooming or sniffing (Gaston et al., 2014; Getsy et al., 2014; Palmer et al., 2013a,b).

HXC AND RETURN TO ROOM-AIR PHASES - FREQ, TV AND MV

Selected examples of the ventilatory waveforms recorded in a SHAM and SCGX mouse during each phase of the protocol are displayed in **Figure 23**. A bar denoting 1.25 sec can be seen between the panels. The baseline Freq (pre-HXC) value was 192 breaths/min in the SHAM mouse and 216 breaths/min in the SCGX mouse. The initial response to HXC (at 15 sec) was very similar in size in the SHAM mouse (192 to 240 breaths/min = 48 breaths/min, +25%) and SCGX mouse (216 to 264 breaths/min = 48 breaths/min, +22%). The decline in Freq (roll-off) that occurred at the end of the 5 min HXC was much more pronounced in the SCGX mouse than in the SHAM mouse. The value (HXC at 5 min) in the SHAM mouse was equivalent to the pre-value (192 to 192 breaths/min = 0 breaths/min,

+0%) whereas the value in the SCGX mouse at 5 min HXC was considerably lower than the pre-value (216 to 168 breaths/min = -48 breaths/min, +22%). The initial increase in Freq upon return to room-air (RA at 15 sec) was more pronounced in the SCGX mouse (216 to 504 breaths/min = 288 breaths/min, +133%) than in the SHAM mouse (192 to 216 breaths/min = 24 breaths/min, +13%). Freq values 5 min following return to room-air (RA at 5 min) were similar in the SHAM mouse (192 to 168 breaths/min = -24 breaths/min, -13%) mouse and SCGX mouse (216 to 192 breaths/min = -24 breaths/min, -11%).

The Freq, TV and MV responses in SHAM and SCGX mice recorded during the 5 min HXC and then return to room-air are shown in **Figure 24**. The left panel in the top row shows that HXC elicited a prompt rise in Freq in SHAM and SCGX mice that decayed substantially (i.e., displayed roll-off) toward the end of the HXC. The middle panel in the top row shows that the initial rise in Freq in SCGX mice was similar to that of SHAM mice except that the response was higher in SCGX mice at 45 sec. The right panel in the top row shows the total increases in Freq in the 0-105 sec, 106-300 sec and 0-300 sec epochs in the SHAM and SCGX mice. As can be seen, the faster roll-off evident in the left-panel resulted in SCGX mice displaying much smaller total response during the 106-300 sec epoch. The return to room-air elicited immediate and pronounced increases in Freq that were similar in SHAM and SCGX mice (top row, left panel). As summarized in **Table 7**, both groups displayed similar total responses if all of the mice were used in the analyses. The relatively large standard error term in the SHAM mice prompted us to examine the data more closely, and we found that the SHAM mice separate into a larger sub-group of 10 that had a net positive change in Freq over the 15 min room-air recording period (+Freq) whereas 4 mice had a net negative response (-Freq). There were 13 mice that fell into the net positive responses in the SCGX mice and only one that fell into the net negative category. The total Freq responses in the SCGX mice that fell into the net positive category were substantially smaller than those in this category of SHAM mice (**Table 7**).

As summarized in the middle row of panels in **Figure 24**, HXC elicited prompt and sustained increases in TV that were very similar in the SHAM and SCGX mice (left and right panels), except that the initial rise in TV happened quicker in SCGX mice, and this rise in TV was significantly higher compared to SHAM mice at 30 sec (middle panel). The return to room-air elicited a minor rise in TV after which TV gradually declined toward pre-HXC values in SHAM and SCGX mice (left panel). As summarized in **Table 7**, both groups had similar total responses. The left panel of the bottom row shows that HXC elicited an initial increase in MV in SHAM and SCGX mice that displayed a marked degree of roll-off. The middle panel shows that the initial responses in the SCGX mice happened somewhat more quickly than in the SHAM mice, and that this augmentation in MV was significantly higher than SHAM mice at 45 sec. The right panel shows that the exaggerated roll-off in the SCGX mice resulted in a smaller total increase in MV during epoch 106-300 sec. As seen in the bottom left panel, there were substantial increases in MV in the SHAM mice and SCGX mice following the return of the room-air. As summarized in **Table 7**, the total MV responses in the SHAM and SCGX groups were similar to one another.

HXC AND RETURN TO ROOM-AIR PHASES – TI AND TE

Figure 25 summarizes the Ti and Te responses in SHAM and SCGX mice during the 5 min HXC and following reintroduction of room-air. As displayed in the panels of the top row, exposure to HXC elicited prompt decreases in Ti (left panel) that occurred at equal rates (middle panel) in the SHAM and SCGX mice. The left panel of the top row shows that the decreases in Ti were subject to roll-off in the latter half of the HXC and that Ti fell to baseline values more quickly SCGX mice. The right panel shows that the total decreases in Ti were smaller in SCGX mice than SHAM mice during the 106-300 sec epoch due to the exaggerated roll-off. As summarized in the bottom left panel, the decreases in Te during the HXC occurred somewhat faster in the SCGX mice and were also subject to roll-off in the SHAM and SCGX mice. This roll-off actually became an increase in Te during the latter stage of HXC in SCGX mice (left panel). As a result, there was a considerable difference in the Te responses of SCGX mice (net increase) and SHAM mice (net decrease) during epoch 106-300 sec. As summarized in **Table 7,** the net changes in Ti and Te, again those displaying positive (+Ti and +Te) and negative (-Ti and -Te) responses, were equivalent in SCGX and SHAM mice following return to room-air.

HXC AND RETURN TO ROOM-AIR PHASES - TI/TE AND INSPIRATORY QUOTIENT

Figure 26 summarizes the Ti/Te and inspiratory quotient (Ti/(Ti+Te)) responses in SHAM and SCGX mice during the 5 min HXC and following re-exposure to room-air. The Ti and Te responses observed during HXC (**Figure 24**) resulted in minor adjustments in Ti/Te and inspiratory quotients in SCGX and SHAM mice that were of similar total magnitude (top and bottom right panels). The reductions in total Ti/Te and Ti/(Ti+Te) following reintroduction of room-air were greater in the SCGX mice than SHAM mice (**Table 7**).

HXC AND RETURN TO ROOM-AIR PHASES - EIP AND EEP

Figure 27 summarizes the EIP and EEP responses in SHAM and SCGX mice during the 5 min HXC and following reintroduction to room-air. The top left panel shows that HXC elicited an immediate reduction in EIP in SHAM and SCGX mice. The top middle panel shows that the initial changes were similar in the SHAM and SCGX mice whereas the top middle panel shows that the total responses were smaller in the SCGX mice during the 106-300 sec epoch and the 0-300 sec epoch. The bottom left panel shows that HXC prompted a decrease in EEP of approximately 2 min in duration in SHAM and SCGX mice. As seen in the bottom panel, EEP went above baseline during the rest of the HXC in the SCGX mice whereas it returned to baseline values in the SHAM mice. As seen in the bottom row middle panel, the reduction in EEP occurred faster in SCGX mice as evident at 15 sec. The bottom right panel shows that the total increases in EEP recorded over the 106-300 sec and 0-300 sec epochs were larger in the SCGX mice than SHAM mice. As summarized in top and bottom left panels, EIP and EEP gradually returned toward baseline values upon return to room-air. **Table 7** shows that the total responses in the SHAM and SCGX mice were similar to one another.

HXC AND RETURN TO ROOM-AIR PHASES - INSPIRATORY DRIVE AND EXPIRATORY DRIVE

Figure 28 summarizes the inspiratory drive (TV/Ti) and expiratory drive (TV/Te) responses in SHAM and SCGX mice during the 5 min HXC and following reintroduction to room-air. The top and bottom left panels show that HXC elicited prompt increases in inspiratory and expiratory drives that displayed roll-off. The top and bottom middle panels show that the initial increases in inspiratory and expiratory drives occurred more quickly in SCGX mice. The top and bottom left panels show that the roll-off was more pronounced in SCGX mice than in SHAM mice such that the total responses were much smaller in the SCGX mice during epoch 106-300 sec panels. The top left panel shows that there was an initial increase in inspiratory drive in the SHAM and SCGX mice upon return to room-air that gradually subsided over time. **Table 7** shows that the total changes in inspiratory drive in SHAM and SCGX mice were similar to one another. The bottom left panel shows there was an initial increase in expiratory drive upon return to room-air that gradually subsided over time. **Table 7** shows that the total changes in expiratory drive were similar in the SHAM and SCGX mice if all mice were considered. However, Table 2 shows that if the SHAM and SCGX mice were divided into two sub-groups, the total changes in expiratory drive were much smaller (i.e., returned to baseline more quickly) in the SCGX mice when only the positive responses were considered.

HXC AND RETURN TO ROOM-AIR PHASES – PIF AND PEF

Figure 29 summarizes the PIF, PEF and PIF/PEF responses in the SHAM and SCGX mice during the 5 min HXC and upon reintroduction of room-air. The top left panel demonstrates that HXC elicited immediate elevations in PIF in SHAM and SCGX mice that displayed a distinct roll-off. The top middle panel shoes that the initial increase in PIF was similar in the SHAM and SCGX mice whereas the top right panel shows that roll-off was considerable faster in SCGX mice such that the total response during the 106-300 sec epoch was much smaller in the SCGX mice than SHAM mice. The middle left panel shows that the increases in PEF occurred faster in SCGX mice whereas the middle middle panel shows that the increase in PEF was significantly faster in the SCGX mice than SHAM mice at 45 sec. The middle right panel shows that he total responses were significantly smaller in the SCGX mice than in the SHAM mice during the 106-300 sec epoch. The top, middle and bottom left panels show that the return to room-air caused initial increases in PIF and PEF in SHAM mice and SCGX mice, and that the degree of these changes resulted in increases in PIF/PEF ratios. **Table 7** shows that total changes in PIF, PEF and PIF/PEF responses were equivalent in SHAM and SCGX mice when all of the mice were included in the analyses. **Table 7** shows that the PIF/PEF ratio was substantially higher if only the positive responses (+PIF/+PEF) were considered.

HXC AND RETURN TO ROOM-AIR PHASES – EF_{50}, R_{PEF} AND RT

Figure 30 summarizes the EF_{50}, Rpef and RT responses in the SHAM and SCGX mice during the 5 min HXC and following return to room-air. As seen in the top left and middle panels HXC caused a brief increase in EF_{50} that happened more quickly in SCGX mice than in SHAM mice and the top middle panel shows that the increase in EF_{50} was significantly faster in the SCGX mice at 45 and 60 sec. The top right panel shows that the total responses were smaller in SCGX mice than SHAM mice during the 106-300 sec epoch. The middle left panel shows that there were initial brief increases in Rpef in SHAM and SCGX mice on initiation of HXC that were followed by sustained reductions in both groups. The middle right panel shows that this pattern of responses was similar in the SHAM and SCGX mice although the reduction in Rpef that occurred in the SHAM mice during the 106-300 sec epoch were not observed in SCGX mice. The bottom middle panel shows that the brief reductions in RT that occurred upon exposure to HXC happened somewhat more rapidly in the SCGX mice than the SHAM mice at 45 sec. The bottom left panel shows that the brief decrease in RT was followed by substantial increases in RT during the HXC in the SCGX mice such that the total responses the occurred in the 106-300 sec epoch was substantially greater in SCGX mice than SHAM mice. The top, middle and bottom left panels show that restoration of room-air was accompanied by the gradual return of EF_{50}, Rpef and RT toward pre-HXC in SHAM and SCGX mice. **Table 7** shows that there were no differences in EF_{50}, Rpef and RT responses between SHAM and SCGX mice when all of the mice were included in the analyses. However, if only positive responses for RT (+RT) and Rpef (+Rpef) were considered (**Table 7**), then the responses were greater in SCGX mice than SHAM mice.

HXC AND RETURN TO ROOM-AIR PHASES – RI, RI/FREQ, APNEIC PAUSES

Figure 31 summarizes the RI, RI/Freq and Apneic Pause responses that occurred in the SHAM and SCGX mice during the 5 min HXC and following restoration of room-air. The top and middle left panels show that HXC elicited prompt elevations in RI of short duration and variable changes in RI/Freq. The top and middle right show that the RI and Freq responses were similar in SHAM and SCGX mice although change in RI, but not RI corrected for Freq (RI/Freq), was less in the SCGX mice between 106-300 sec. The bottom left panel shows that the degree of apneic pauses decreased during the HXC in SHAM and SCGX mice, and the bottom right panel shows that the total reduction was more pronounced in SCGX mice. The top, middle and bottom left panels show that restoration of room-air was associated with pronounced and rapid increases in RI, RI/Freq and the degree of apneic pauses. As shown in **Table 7**, the total magnitude of these responses in SHAM and SCGX mice were similar to each other when all mice were included in the analyses. Additionally, as seen in **Table 7**, when only negative apneic pauses (-Te/RT)-1) were considered, it was evident that the SCGX mice showed a greater reduction in apneic pauses (i.e., recovered more quickly) than SHAM mice.

DISCUSSION

The present study demonstrates that resting ventilatory parameters were equivalent in the SCGX and SHAM mice except for a greater incidence of apneic pauses in SCGX mice. Taken together, this data indicates that the absence of SCG projections to structures controlling breathing including brainstem nuclei, upper airway and carotid bodies, (Getsy et al., 2021), does not dramatically affect baseline ventilatory timing and/or mechanics, whereas it can affect the quality of breathing by enhancing the incidence of non-eupneic breaths (i.e., apneic pauses). The finding that resting RI and RI/Freq in SCGX mice was not different from SHAM mice suggests that the incidence of central apneas and sighs (Getsy et al., 2014) was not affected by the removal of SCG input to central and/or peripheral structures. However, the enhanced degree of apneic pauses raises the possibility that the loss of SCG projections to the upper airway (e.g., nasopharynx, glottus, epiglottus,) and tongue has subtle effects on the functional mechanics of these structures, although it is perhaps more likely to be due to the loss of glossopharyngeal sensory innervation of SIF cells (Takaki et al., 2015). The increased incidence of apneic pauses in SCGX mice was not observed in CSCX mice (Getsy et al., 2021), which suggests that the direct loss of SCG projections is not equivalent to cutting the preganglionic input to the SCG. It is expected that transection of the CSC would result in alterations in cellular processes within the SIF cells and postganglionic neurons of the SCG that, in turn, could lead to changes in firing properties of these interneurons and neurons, respectfully. However, we could find no published literature for the possibility that damage to the preganglionic motor neurons within the T1-T4 spinal cord or their fibers within the CSC-SCG pathways would do anything to diminish the activities of postganglionic SCG neurons and SIF cells of the SCG. As such the question remains as to the reasons why CSCX and SCGX have different effects on resting breathing. Numerous studies (McLachlan, 2007; Qu et al., 1988; Stein and Weaver, 1988; Meckler and Weaver, 1985) have provided evidence that the loss of preganglionic input elicits sustained decreases in postganglionic nerve activity in sympathetic ganglia. In contrast, Qu and colleagues found that firing of splenic and mesenteric postganglionic nerves continues unabated after damage to the spinal cord despite substantially diminished preganglionic outflow to splenic and inferior mesenteric ganglia (Qu et al., 1988). Therefore, the likelihood that postganglionic neurons and SIF cells in the SCG continue to fire after transection of the CSC remains a possibility. It must be remembered that these SCGX studies were performed only 4 days after surgical removal of the SCG, and it is certainly a possibility that more exaggerated changes in resting breathing would occur at times greater than 4 days post-SCGX.

In agreement with Getsy et al. (2021), the present study shows that (a) HXC elicited an initial increase in Freq in the SHAM C57BL6 mice that was associated with decreases in Ti, Te, EEP, Rpef and RT, (b) all of these responses were subject to substantial roll-off, (c) the pronounced reductions in EIP were sustained throughout HXC, and (d) HXC elicited substantial and sustained elevations in inspiratory and expiratory drives as we as

TV, MV, PIF, PEF and EF_{50}. The description of the C57BL6 mice with respect to ventilatory control processes, and potential answers as to why some ventilatory parameters in C57BL6 mice express roll-off whereas other parameters do not was discussed by Getsy et al. (2021). While it is evident that the patterns and levels of non-eupneic breathing in C57BL6 mice and their ventilatory responses to HXC have a genetic component, there is no mechanistic evidence as to why ventilatory parameters in these mice do or do not display roll-off. Indeed, ventilatory mechanics such as Rpef, and RT and ventilatory timing parameters such as Freq and EEP display roll-off during HXC whereas mechanics parameters such as PIF, PEF and EF_{50} and timing parameters such as EIP do not display roll-off. Coming to terms with how each of these ventilatory responses contributes to the overall HXC response will help to better define the physiological processes recruited by HXC, and perhaps help to identify the potential processes by which disease states elicit breathing disorders and abnormal responses to HXC.

In addition to studying resting ventilatory parameters, we show that freely-moving C57BL6 male mice with SCGX displayed numerous altered responses to HXC that also occurred in C57BL6 male mice with CSCX (e.g., more rapid occurrence of changes in frequency of breathing, minute ventilation, expiratory drive and EF_{50}) (data for CSCX study is in our companion manuscript Getsy et al., 2021). However, as detailed in **Table 8**, the SCGX mice also displayed numerous changes in HXC responses that CSCX mice did not, including reductions in the total increases in Freq, MV, PIF, PEF, inspiratory and expiratory drives and RI (i.e., occurrence of non-eupneic breaths). Taken together with the findings of Getsy et al. (2021), it is evident that bilateral CSCX is not equivalent to bilateral SCGX, which suggests that hypoxia has a multiplicity of important effects on neural signaling within the CSC-SCG signaling pathways. The increase in RI at the onset of HXC and upon return to room-air is a unique feature of C57BL6 mice but not Swiss Webster mice or B6AF1 (C57BL6 dam × A/J sire) mice, which display minimal changes in RI upon return to room-air (Getsy et al., 2014). We do not know the exact mechanisms by which RI changes so remarkably in C57BL6 mice, but conjecture that processes within carotid body glomus cells, carotid sinus nerve chemoafferents and/or neurons within the NTS that process chemoafferent information, allow for exaggerated expression of non-eupneic breathing at the onset of HXC and upon return to room-air

These data raise the possibility that HXC may directly activate sub-populations cells in the SCG, such as SIF cells and postganglionic neurons, independently of increased input from the CSC, and that this SCG input to cardiorespiratory structures has dual roles of dampening the initial occurrence of the hypoxic ventilatory response, while promoting the overall magnitude of the response. These apparently divergent effects may be due to the loss of input to peripheral and central structures with differential roles in the control of breathing in response to HXC. The interpretation of our findings must include the compelling evidence that sub-populations of SCG cells and especially SIF cells are hypoxia-sensitive (Nunes et al., 2010, 2012; Strosznajder et al., 1997; Dinger et al., 1993; Brokaw and Hansen, 1987; Hanson et al., 1986) although there is equally compelling evidence that the SCG principal neurons themselves are not directly sensitive to hypoxia (Bernardini et al., 2020; Gao et al., 2019; Buckler and Turner, 2013; Nunes et al., 2012; Rigual et al., 1999).

The postganglionic cells in the SCG project to a numerous structures that participate in the regulation of ventilation including, nuclei in the brainstem (e.g., NTS) and hypothalamus (Mathew, 2007; Wiberg and Widenfalk, 1993; Hughes-Davis et al., 2005; Esquifino et al., 2004; Westerhaus and Loewy, 1999; Saavedra, 1985; Gallardo et al., 1984; Cardinali et al., 1981a,b, 1982) and peripheral structures such as the upper airway, tongue and carotid body (O'Halloran et al., 1996, 1998; Hisa et al., 1999; Kummer et al., 1992; Oh et al., 2006; Wang and Chiou, 2004; Flett and Bell, 1991). There is substantial evidence that the SCG has important roles in the integration of the ventilatory and cardiovascular systems including that activation of the CSC in rats elicits profound decreases upper airway resistance and arterial blood pressure (O'Halloran et al., 1996, 1998). As discussed by Prabhakar (1994), there is no clear agreement about the ability of the sympathetic nerve supply to the carotid bodies to influence the resting (normoxic) activity of type 1 glomus cells and/or chemoreceptor afferent nerve fibers and how this input modulates the responses of these structures during hypoxic exposure. For instance, in cats, CSC and GGN activity increases during exposure to hypoxic challenges (Yokoyama et al., 2015; Lahiri et al., 1986; Matsumoto et al., 1986, 1987), and that activation of the GGN input to the carotid bodies decreases the hypoxic responsiveness of the primary chemosensors (McQueen et el., 1989). Moreover, diverse responses have been reported foowing application of sympathetic neurotransmitters such as norepinephrine and dopamine to *in vitro* and *in vivo* carotid body preparations, including (1) direct inhibitory effects on glomus cell and/or chemoreceptor afferent activity (Overholt and Prabhakar, 1999; Almaraz, et al., 1997; Ryan et al., 1995; Prabhakar et al., 1993; Bisgard et al., 1993;

Pizarro et al., 1992; Kou et al., 1991; Folgering et al., 1982; Mills et al., 1978; Llados and Zapata, 1978; Zapata, 1975; Zapata et al., 1969), (2) direct activation of glomus cells and/or chemoafferent fibers (Pang et al., 1999), Heinert et al., 1995; Milsom et al., 1983; Lahiri et al., 1981; Matsumoto et al., 1980) (3) indirect excitation of primary glomus cells within the carotid body via constriction of microvascuar blood vessels within the carotid body (Yokoyama et al., 2015; Potter et al., 1987), (4) a biphasic response pattern that consists of initial brief reduction in carotid sinus nerve activity followed by a longer-lasting period excitation (Matsumoto et al., 1981), and (5) a biphasic response pattern that consists of initial brief burst in carotid sinus nerve activity followed by long-lasting inhibition (Bisgard et al., 1979). However, it must be remembered that the sympathetic neurotransmitters listed above are also present in sub-populations of carotid body glomus cells and therefore can act in an autocrine-paracrine manner without postganglionic sympathetic involvement. In the context of sympathetic innervation of glomus cells, this is a confounding variable with respect to understanding the interplay between the sympathetic nerves and glomus cells during a hypoxic challenge. At present, we do not know (1) which nerve fibers (e.g., ICN, ECN and/or GGN) postganglionic SCG neurons travel with to modulate the HXC-induced ventilatory responses or what the exact targets are, or (2) the exact nerve tracts and central projections of sensory afferents emanating from the SCG (Brognara et al., 2020; Takaki et al., 2015). We are in the process of determining the effects of resulting from transection of postganglionic sympathetic trunks (e.g., GGN, ECN, and ICN) to delineate the functional roles of postganglionic sympathetic innervation to the carotid bodies (i.e., studies involving GGNX) as opposed to other central and peripheral structures (i.e., studies involving ICNX and/or ECNX).

VENTILATORY RESPONSES THAT OCCURRED FOLLOWING RETURN TO ROOM-AIR

Previous studies have found that the restoration of room-air following HXC can result in distinct respiratory patterns that can be categorized as post-hypoxic frequency decline, when Freq falls below baseline levels (Dick and Coles, 2000) or short-term potentiation, when ventilation stays higher than baseline values for a considerable amount of time (Getsy et al., 2014; Powell et al., 1998;). The C57BL6 mice used in this study and that of Getsy et al. (2021) showed substantially elevated Freq following reintroduction of room-air that was associated with a prolonged phase of non-eupneic (disordered) breathing. The potential mechanisms that underlie post-HXC disordered breathing include disturbances in brainstem signaling pathways (Strohl, 2003; Dick and Coles, 2000; Wilkinson et al., 1997; Coles and Dick, 1996) but not altered signaling events within the carotid bodies (Brown et al., 1993; Vizek et al., 1987) although there is compelling evidence that disturbances in carotid body chemoreceptor afferent signaling have important roles in the etiology of sleep apneas (Smith et al., 2003). It should be noted that the post-HXC responses in the SHAM mice often showed considerable variability that prompted us to re-analyze the data and separate the mice into 2 sub-groups - those that displayed net-positive responses (e.g., mice that had an overall increase in Freq during the 15 min recording period) and those that displayed net-negative responses (e.g., mice that had an overall decrease in Freq during the 15 min recording period). Since all mice appeared healthy in every way, it would seem that this variability is related to the mixed genetics of the C57BL6 mice (Campen et al., 2004; Tankersley, 2003, 2001; Tankersley et al., 2002, 2000, 1994). When all 14 SHAM mice and 14 SCGX mice were compared, we found that (1) the negative Ti/Te and Ti/(Ti+Te) responses were more pronounced in SCGX mice (largely because of differences in the changes in Te) (**Table 7**), (2) RT minimally changed in SHAM mice, but significantly increased in SCGX mice (**Table 7**). However, when taking the dominant positive or negative responses into consideration, we find that (1) the total increases in Freq, and expiratory drive were significantly smaller in SCGX mice, (2) the total increases in PIF/PEF ratios and Rpef were greater in SCGX mice, and (3) the total decreases in apneic pauses were greater in the SCGX mice. Therefore, it appears that the absence of SCG input to ventilatory control systems has significant consequences for parameters involved in ventilatory timing and mechanics.

POTENTIAL ROLES OF SMALL INTENSELY FLUORESCENT (SIF) CELLS IN THE SCG

Small intensely fluorescent (SIF) cells comprise only about 3% of the total SCG cell population and were initially described as dopaminergic interneurons that received innervation from preganglionic cholinergic fibers within the CSC (Libet, 1970; Eccles and Libet, 1961). Whereas principal SCG neurons are the main source of norepinephrine in the SCG, the SIF cells are proportionately the richest source of dopamine and store about 50% of the total amount of dopamine in the SCG (Soulier et al., 1997; Borghini et al., 1991). Moreover, Libet and Owman (1974) reported that dopamine released from SIF cells mediated slow inhibition of principal SCG neurons (i.e., dopamine generated slow inhibitory postsynaptic potential), thereby modulating postganglionic output from the SCG. SIF cells show morphological and biochemical similarities to hypoxia-sensitive carotid body type I (glomus) cells (Dalmaz et al., 1993) and reside in clusters near fenestrated capillaries like carotid body glomus cells are in clusters located next to fenestrated capillaries (McDonald and Blewett, 1981). Unlike principal postganglionic SCG neurons, SCG-SIF cells possess O_2-sensing properties (Dinger et al., 1993 and references below). Although direct evidence is lacking, there is strong other support in favor of SIF cells being the hypoxia-sensitive cell type in the SCG. This evidence includes that (1) the morphological plasticity following prolonged hypoxia is similar in dopaminergic carotid body glomus cells and SCG-SIF cells (Soulier et al. 1997; Dalmaz et al., 1993), (2) similar to carotid body glomus cells, SIF cells receive both afferent and efferent innervation (McDonald, 1983a,b; Zaidi and Matthews, 2013; Takaki et al. 2015), (3) a combination of immunohistochemistry and retrograde labeling has allowed identification of three types of SIF cells in SCG and again similar to carotid body glomus cells, most afferent endings on SIF cells originated from the petrosal ganglion (Takaki et al., 2015), and (4) afferent endings on SIF cells express purinergic P2X3 receptors (Takaki et al., 2015), thus revealing a sensory pathway analogous to carotid body chemoreceptors where ATP acting on P2X3-containing postsynaptic receptors on petrosal terminals is a major mechanism for hypoxic chemo-transmission (Nurse, 2014). Taken together, the possibility arises that the ability of SCGX (removal of all input/output signals) to affect the hypoxic ventilatory responses in a fashion clearly different from CSCX (removal of preganglionic input) may arise from (1) the ability of SIF cells to be directly activated by HX challenges independently of CSC input, and (2) the ability of SCGX to remove the SIF cell-sensory neuron interactions within the SCG. It remains to be determined whether our findings are relevant to other species given the broad species-dependent variability in the sympathetic supply to the carotid

vasculature and the resulting control of carotid body sensitivity (Brognara et al., 2020)

CONCLUSION

The data presented in this manuscript provide compelling evidence that the CSC-SCG complex of C57BL6 mice may provide inhibitory input to central and peripheral structures that participate in expression of the initial ventilatory responses to HXC. However, the present study demonstrates that bilateral SCGX has greater effects on the HXC-induced responses than CSCX (Getsy et al., 2021. As discussed above, this raises the possibility that HXC may directly alter the activity of postganglionic neurons in the SCG independently of input from the ipsilateral CSC. If the SIF cells rather than principal SCG neurons are indeed the direct hypoxia-sensitive targets (as would seem more likely), then their afferent connections via the glossopharyngeal nerve (Takaki et al., 2015) may be sufficient to drive the chemoreflex response without principal SCG neuron involvement or CSC input. Moreover, the dopaminergic inhibitory pathway from SIF cells to principal SCG neurons could still be active during direct hypoxic stimulation of SIF cells and this could further modulate postganglionic sympathetic input to the carotid body (Tosaka and Kobayashi, 1977; Libet and Tosaka, 1970). A related key question is whether or not SIF cells in mouse SCG also receive a glossopharyngeal P2X3-mediated sensory innervation as reported for rat SCG (Takaki et al., 2015), and whether this pathway is intact and functional following CSCX. Assuming that some SIF cells in mouse SCG receive cholinergic CSC and glossopharyngeal P2X3-afferent innervation as seen in rat SCG (Takaki et al. 2015), the presence of this parallel chemoafferent reflex circuit may be sufficient to unify the present findings with those of our companion CSCX study (Getsy et al., 2021). Thus, CSCX would selectively eliminate the cholinergic innervation of O_2-chemosensitive SIF cells and thereby modulate the afferent sensory output via this SIF-glossopharyngeal circuit during hypoxia. On the other hand, SCGX would eliminate both cholinergic input to SIF cells and glossopharyngeal sensory output during hypoxia. Therefore, the activity of the CSC-SIF cell-glossopharyngeal circuit alone may account for the observed differences in respiratory parameters following CSCX and SCGX. The SCG projects widely to both central and peripheral structures that control breathing, and we are performing studies to determine the effects of transection of the major postganglionic branches of the SCG (i.e., ICN, ECN and GGN) on ventilatory function in mice and their responses to HXC. In addition, to determine the genetic and temporal aspects of SCGX in the control of ventilation and the responses to HXC we are performing experiments using C57BL6, Swiss-Webster and A/J mice (Getsy et al., 2014) studied 4, 14, 28 and 56 days post-SCGX.

Table 6

Baseline parameters in sham-operated (SHAM) mice and in mice with bilateral removal of the superior cervical ganglia (SCGX)

Parameter	SHAM	SCGX
Number of mice	14	14
Age, days	99 ± 1	98 ± 1
Body Weight, grams	27.6 ± 0.5	28.1 ± 0.5
Frequency (Freq, breaths/min)	194 ± 5	190 ± 4
Tidal Volume (TV, ml)	0.151 ± 0.004	0.150 ± 0.006
Minute Ventilation (ml/min)	28.7 ± 0.7	28.2 ± 0.8
Inspiratory Time (Ti, sec)	0.114 ± 0.002	0.115 ± 0.002
Expiratory Time (Te, sec)	0.212 ± 0.008	0.216 ± 0.005
End Inspiratory Pause (EIP, msec)	2.89 ± 0.09	2.92 ± 0.07
End Expiratory Pause (EEP, msec)	47.5 ± 11.6	42.9 ± 7.7
Ti/Te	0.555 ± 0.018	0.542 ± 0.012
Ti/(Ti + Te)	0.354 ± 0.008	0.350 ± 0.005
Peak Inspiratory Flow (PIF, ml/sec)	2.34 ± 0.05	2.31 ± 0.04
Peak Inspiratory Flow (PIF, ml/sec)	1.58 ± 0.06	1.60 ± 0.09
PIF/PEF	1.51 ± 0.06	1.50 ± 0.07
Expiratory flow at 50% exhaled TV (EF$_{50}$, ml/sec)	0.072 ± 0.002	0.073 ± 0.002
Relaxation Time (sec)	0.099 ± 0.003	0.093 ± 0.002
Rate of achieving PEF (Rpef)	0.141 ± 0.009	0.127 ± 0.009
Inspiratory Drive (TV/Ti, ml/sec)	1.34 ± 0.03	1.32 ± 0.03
Expiratory Drive (TV/Te, ml/sec)	0.73 ± 0.03	0.71 ± 0.03
Rejection Index (%)	15.0 ± 2.0	14.1 ± 1.6
(Rejection Index/Frequency) x 100	8.1 ± 1.2	7.5 ± 0.9
Apneic Pauses	1.15 ± 0.06	1.33 ± 0.04*

The data are presented as mean $\pm$ SEM. *P < 0.05, SCGX *versus* SHAM.

Table 7

Total changes that occurred during the first 15 min upon return to room-air

Parameter	SHAM	SCGX
Number of mice	14	14
Frequency (%)	+41.1 ± 10.8 (14)	+30.9 ± 5.5 (14)
+ Freq, %	+58.8. ± 8.9 (10)	+33.5 ± 5.0 (13)*
- Freq, %	-3.3 ± 0.5 (4)	-2.9 (1)
Tidal Volume (TV, %)	+39.2 ± 8.0 (14)	+28.0 ± 6.3 (14)
+ TV, %	+43.4 ± 7.1 (13)	+33.3 ± 5.6 (12)
- TV, %	-15.8 (1)	-2.9 ± 0.1 (2)
Minute Ventilation (%)	+113 ± 26.9	+79.7 ± 14.1 (14)
+ MV, %	+123.2 ± 25.9 (13)	+79.7 ± 14.1 (14)
- MV, %	-20 (1)	none
Inspiratory Time (Ti, %)	-30.4 ± 5.3 (14)	-30.3 ± 3.1 (14)
Expiratory Time (Te, %)	-4.1 ± 6.0 (14)	+8.6 ± 4.9 (14)
+ Te,%	-19.5 ± 4.2 (8)	-12.3 ± 2.3 (4)
- Te, %	+16.5 ± 2.2 (6)	+17.0 ± 3.7 (10)
Ti/Te, %	-25.9 ± 2.3 (14)	-33.3 ± 1.9* (14)
Ti/(Ti + Te), %	-19.0 ± 1.7 (14)	-25.1 ± 1.5* (14)
End Inspiratory Pause (EIP, %)	-8.3 ± 2.7 (14)	-12.1 ± 2.5 (14)
End Expiratory Pause (EEP, msec)	+185 ± 55 (14)	+259 ± 66 (14)
Inspiratory Drive (TV/Ti, %)	+143 ± 29 (14)	+116 ± 16 (14)
Expiratory Drive (TV/Te, %)	+77.8 ± 21.4 (14)	+44.6 ± 11.8 (14)
+ TV/Te,%	+100.9 ± 19.9 (11)	+53.2 ± 11.2* (12)
- TV/Te, %	-7.2 ± 3.6 (3)	-6.9 ± 1.8 (2)

The data are presented as mean ± SEM. *P < 0.05, SCGX versus SHAM. The values in parentheses equal the number of mice in the sub-groups for Te, RT, Rpef and (Te/RT)-1.

Table 7 - continued

Total changes that occurred during the first 15 min upon return to room-air

Parameter	SHAM	SCGX
Number of mice	14	14
Peak Inspiratory Flow (PIF, %)	+136 ± 26 (14)	+113 ± 14 (14)
Peak Expiratory Flow (PEF, %)	+122 ± 18 (14)	+83 ± 16 (14)
PIF/PEF, %	+11.1 ± 5.6 (14)	+25.8 ± 6.3 (14)
+ PIF/PEF,%	+18.2 ± 4.7 (11)	+34.6 ± 4.8* (11)
- PIF/PEF, %	-15.0 ± 1.6 (3)	-6.6 ± 1.5 (3)
Expiratory flow at 50% exhaled TV (EF$_{50}$, %)	+117 ± 27 (14)	+80 ± 17 (14)
Relaxation Time (RT, %)	-5.3 ± 5.5 (14)	+15.0 ± 4.6* (14)
+ RT,%	+12.1 ± 3.3 (8)	+24.5 ± 2.1* (10)
- RT, %	-16.0 ± 3.9 (6)	-8.9 ± 2.1 (4)
Rate of achieving PEF (Rpef, %)	+14.5 ± 10.7 (14)	+19.6 ± 13.9 (14)
+ Rpef,%	+40.4 ± 6.8 (8)	+69.9 ± 9.6* (6)
- Rpef, %	-20.1 ± 3.3 (6)	-18.0 ± 4.8 (8)
Rejection Index (RI, %)	+315 ± 84 (14)	+280 ± 52 (14)
(Rejection Index/Frequency) x 100 (%)	+175 ± 41 (14)	+194 ± 37 (14)
Apneic Pauses (Te/RT)-1 (%)	+5.3 ± 4.8 (14)	-4.9 ± 6.1 (14)
+ Te/RT)-1,%	+19.1 ± 3.7 (7)	+25.9 ± 4.1 (4)
- Te/RT)-1, %	-8.6 ± 1.9 (7)	-17.2 ± 2.3* (10)

The data are presented as mean ± SEM. *P < 0.05, SCGX versus SHAM. The values in parentheses equal the number of mice in the sub-groups for Te, RT, Rpef and (Te/RT)-1

Table 8

A qualitative assessment of the responses in C57BL6 mice with bilateral transection of the cervical sympathetic chain (CSCX) compared to their SHAM controls and in mice with bilateral removal of the superior cervical ganglia (SCGX) compared to their SHAM controls

Parameter	Initial		Epoch	
	CSCX	SCGX	CSCX	SCGX
Increase in Frequency	Faster			↓106-300
Increase in Tidal Volume		Faster	↑106-300	
Increase in Minute Ventilation	Faster	Faster		↓106-300
Decrease in Inspiratory Time (Ti)	Faster			↓106-300
Decrease in Expiratory Time (Te)	Faster	Faster		↓106-300
Decrease in Ti/Te			↓106-300	
Decrease in Ti/(Ti + Te)			↓106-300	
Decrease in End Inspiratory Pause				↓106-300
*Decrease in End Expiratory Pause	Faster	Faster		
*Increase in End Expiratory Pause				↑106-300
Increase in Inspiratory Drive	Faster	Faster		↓106-300
Increase in Expiratory Drive	Faster	Faster		↓106-300
Increase in Peak Inspiratory Flow				↓106-300
Increase in Peak Expiratory Flow	Faster	Faster	↑106-300	↓106-300
Decrease in PIF/PEF	Slower		↑106-300	
*Decrease in Relaxation Time	Faster	Faster		
*Increase in Relaxation Time				↑106-300
Decrease in Rate of achieving PEF			↑106-300	↓106-300
Increase in EF_{50}, ml/sec	Faster	Faster		↓106-300
Increase in Rejection Index	Faster			↓106-300
Increase in Rejection Index/Frequency	Faster	Slower		↓106-300
Decrease in Apneic Pauses (Te/RT)-1				↑106-300

↑ = enhancement compared to respective SHAM group, ↓ = reduction compared to respective SHAM group.

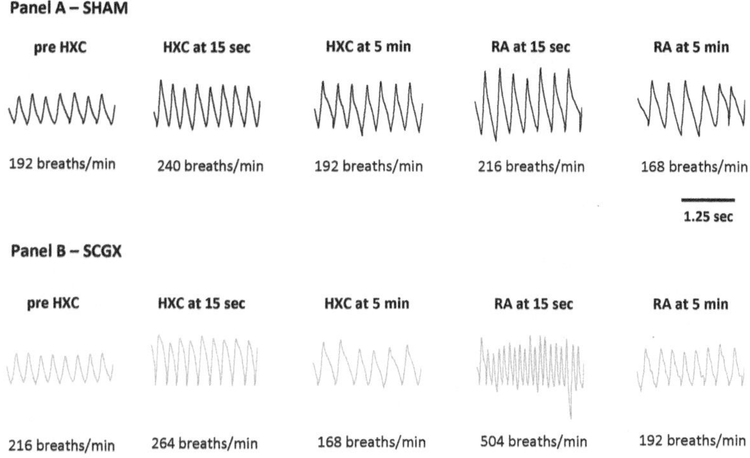

Figure 23. Selected sections of the respiratory waveforms that were recorded during key times of the protocol in a sham-operated (SHAM) mouse and in a mouse with bilateral superior cervical ganglionectomy (SCGX). A 1.25 sec time bar is displayed between the panels. Abbreviations: HXC, phase of challenge with hypoxic gas. RA, the room-air phase.

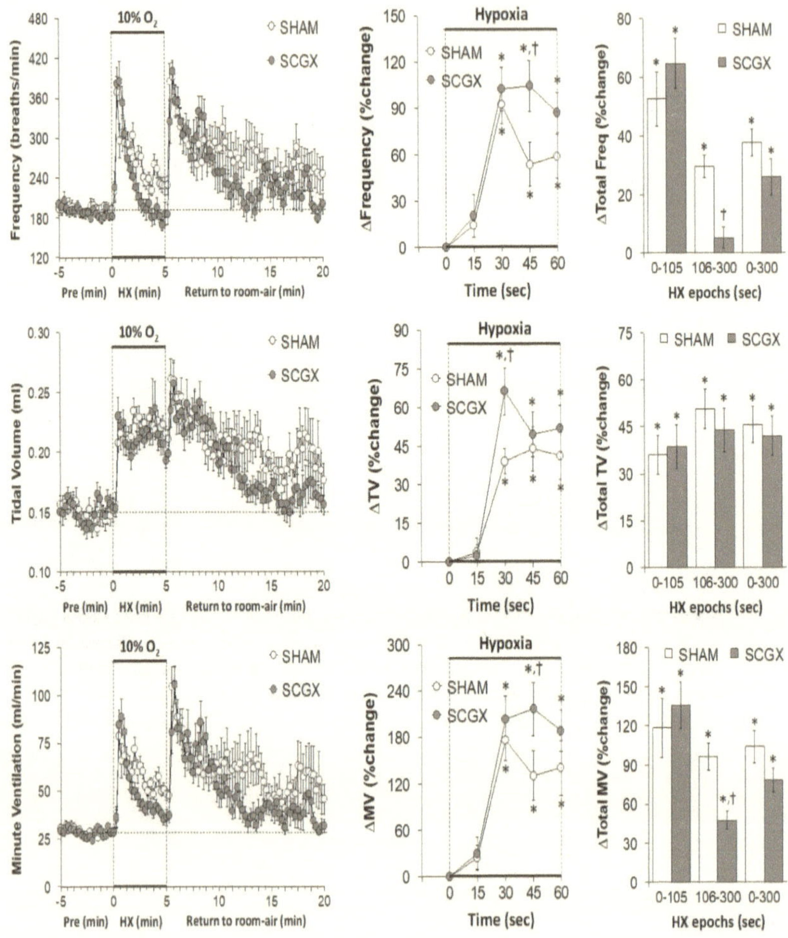

Figure 24. Left panels: Frequency of breathing (Freq), tidal volume (TV) and minute ventilation (MV) values recorded prior to, during and following a hypoxic (HX, 10% O_2, 90% N_2) gas challenge of 5 min in duration in mice that were sham-operated (SHAM) and those with bilateral superior cervical ganglionectomy (SCGX). **Middle Panels:** Frequency of breathing, tidal volume and minute ventilation responses recorded during the first min of HXC in the SHAM and SCGX rats (expressed as % pre-values). **Right panels:** Total responses recorded during the first 105 sec, between 106-300 sec and between 0-300 sec (expressed as the sum of all % changes from pre). All results are presented as mean ± SEM. *P < 0.05, significant change from Pre-values. †P < 0.05, SCGX mice *versus* SHAM mice. There were 14 mice in each of the SHAM and SCGX groups.

132

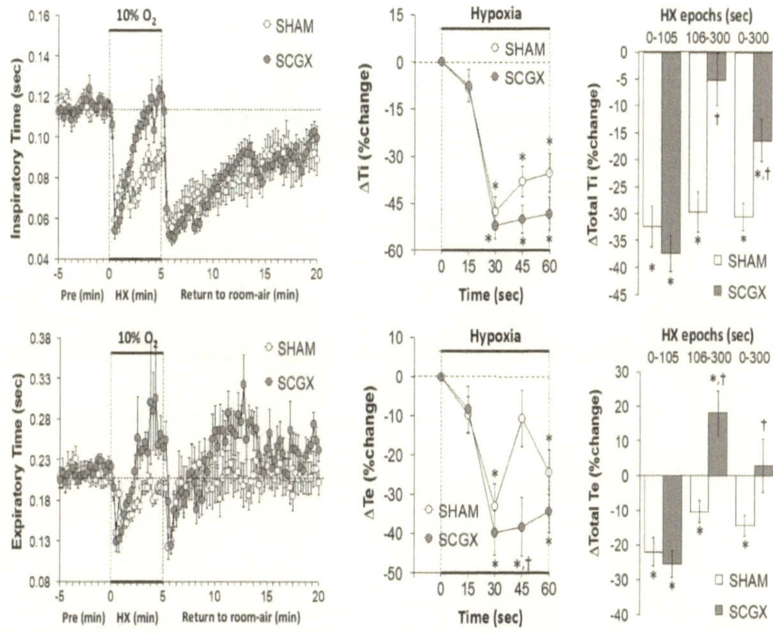

Figure 25. Left panels: Inspiratory time (Ti) and expiratory time (Te) values recorded prior to, during and following a hypoxic (HX, 10% O_2, 90% N_2) gas challenge of 5 min in duration in mice that were sham-operated (SHAM) and those with bilateral superior cervical ganglionectomy (SCGX). **Middle Panels:** Responses of inspiratory times and expiratory times recorded during the first minute of HXC in the SHAM and SCGX rats (expressed as % of pre-values). **Right panels:** Total responses recorded during the first 105 sec, between 106-300 sec and between 0-300 sec (expressed as the sum of all % changes from pre). All results are presented as mean ± SEM. *P < 0.05, significant from pre-values. †P < 0.05, SCGX *versus* SHAM. There were 14 mice in each of the SHAM and SCGX groups.

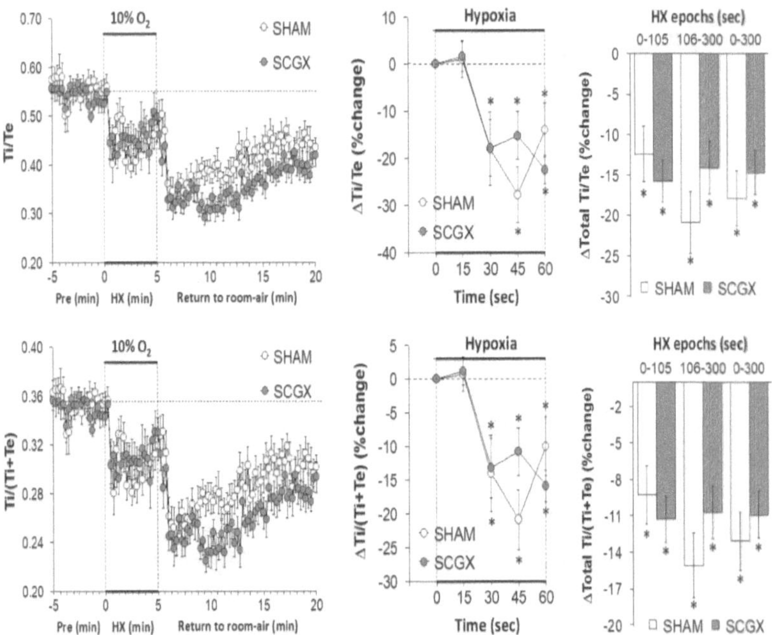

Figure 26. Left panels: Inspiratory time/expiratory time (Ti/Te) ratios and Ti/(Ti+Te) ratios recorded prior to, during and following a hypoxic (HX, 10% O_2, 90% N_2) gas challenge of 5 min in duration in mice that were sham-operated (SHAM) and those with bilateral superior cervical ganglionectomy (SCGX). **Middle Panels:** Responses recorded during the first minute of HXC in the SHAM and SCGX rats (expressed as % of pre-values). **Right panels:** Total responses recorded during the first 105 seconds, between 106-300 sec and between 0-300 sec (expressed as the sum of all % changes from pre). All results are presented as mean ± SEM. *P < 0.05, significant from pre-values. †P < 0.05, SCGX *versus* SHAM. There were 14 mice in each of the SHAM and SCGX groups

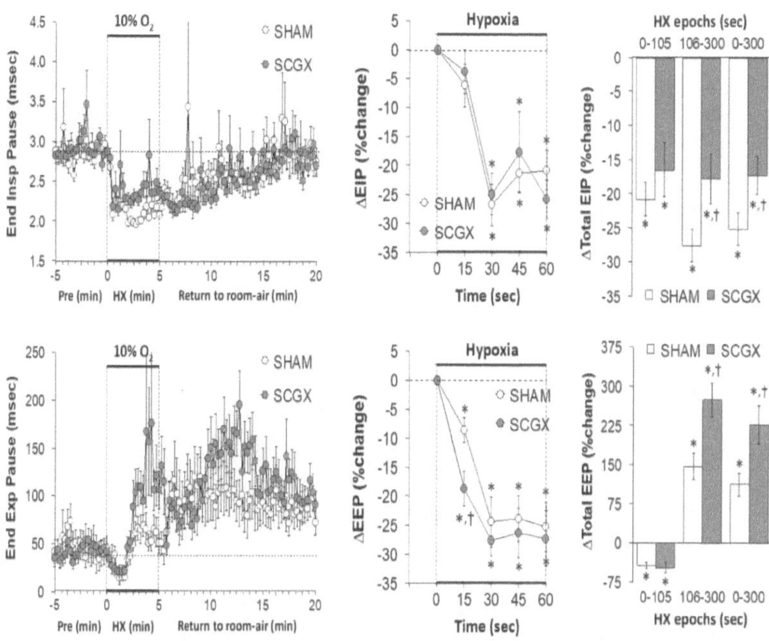

Figure 27. Left panels: End inspiratory pause (EIP) and end expiratory pause (EEP) values recorded prior to during, and following a hypoxic (HX, 10% O_2, 90% N_2) gas challenge of 5 min in duration in mice that were sham-operated (SHAM) and those with bilateral superior cervical ganglionectomy (SCGX). **Middle Panels:** Responses recorded during the first minute of HXC in the SHAM and SCGX rats (expressed as % of pre-values). **Right panels:** Total responses recorded during the first 105 sec, between 106-300 sec, and between 0-300 sec (expressed as the sum of all % changes from pre). All results are presented as mean ± SEM. *P < 0.05, significant change from pre-values. †P < 0.05, SCGX *versus* SHAM. There were 14 mice in each of the SHAM and SCGX groups.

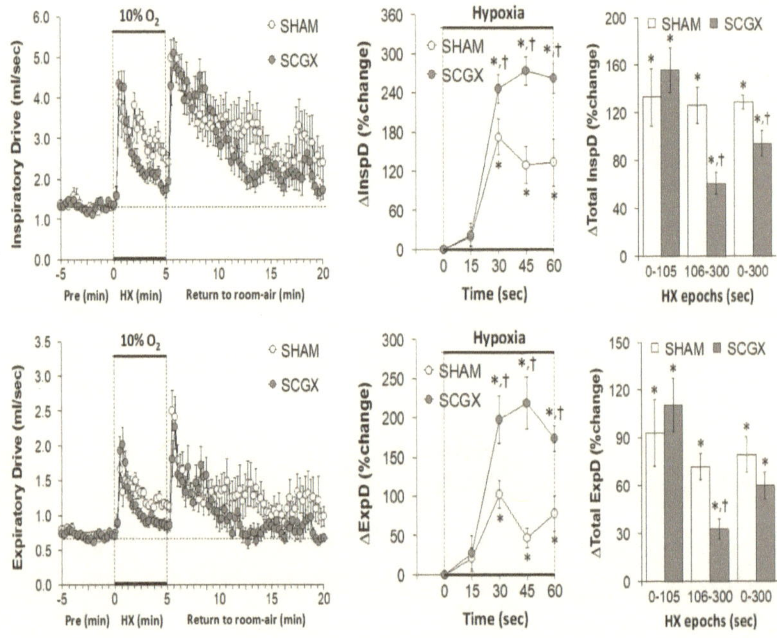

Figure 28. Left panels: Inspiratory Drive (TV/Ti) and Expiratory Drive (TV/Te) values recorded prior to, during, and following a hypoxic (HX, 10% O₂, 90% N₂) gas challenge of 5 min in duration in mice that were sham-operated (SHAM) and those with bilateral superior cervical ganglionectomy (SCGX). **Middle Panels:** Responses recorded during the first minute of HX challenge in the SHAM and SCGX rats (expressed as % of pre-values). **Right panels:** Total responses recorded during the first 105 sec, between 106-300 sec and between 0-300 sec (expressed as the sum of all % changes from pre). All results are presented as mean ± SEM. *P < 0.05, significant change from pre-values. †P < 0.05, SCGX *versus* SHAM. There were 14 mice in each of the SHAM and SCGX groups.

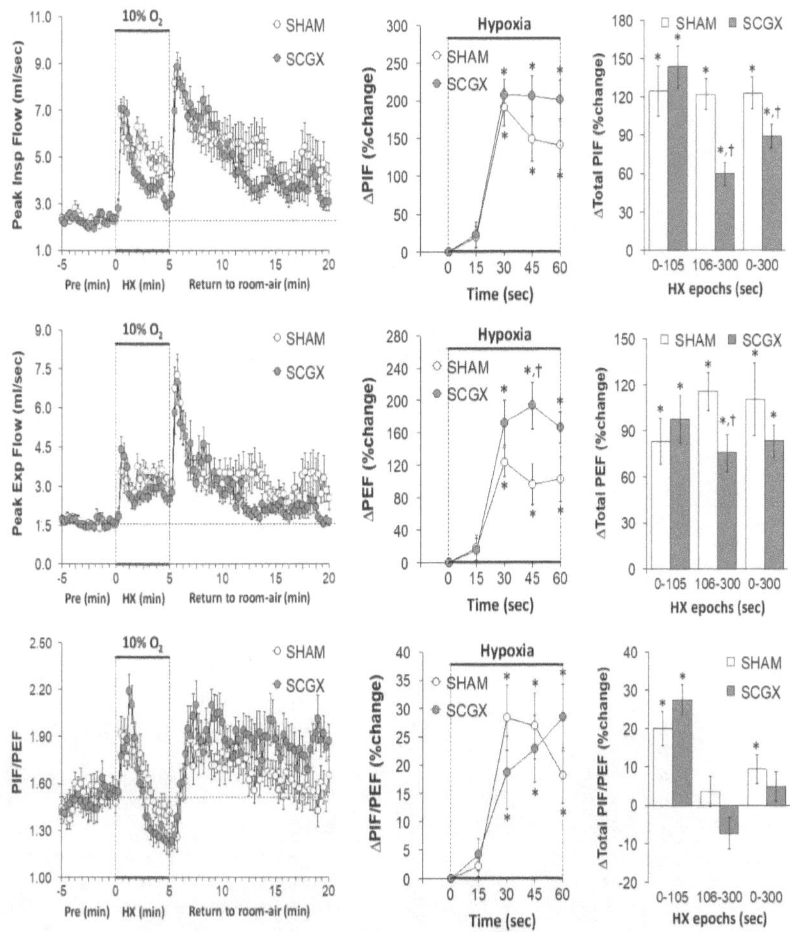

Figure 29. Left panels: Peak Inspiratory flow (PIF) and Peak Expiratory Flow (PEF) values and PIF/PEF ratios recorded prior to, during and following a hypoxic (HX, 10% O_2, 90% N_2) gas challenge of 5 min in duration in mice that were sham-operated (SHAM) and those with bilateral superior cervical ganglionectomy (SCGX). **Middle Panels:** Responses recorded during the first minute of HXC in SHAM and SCGX rats (expressed as % of pre-values). **Right panels:** Total responses recorded during the first 105 seconds, between 106-300 sec and between 0-300 sec (expressed as the sum of all % changes from pre). All results are presented as mean ± SEM. *P < 0.05, significant change from pre-values. †P < 0.05, SCGX *versus* SHAM. There were 14 mice in each of the SHAM and SCGX groups.

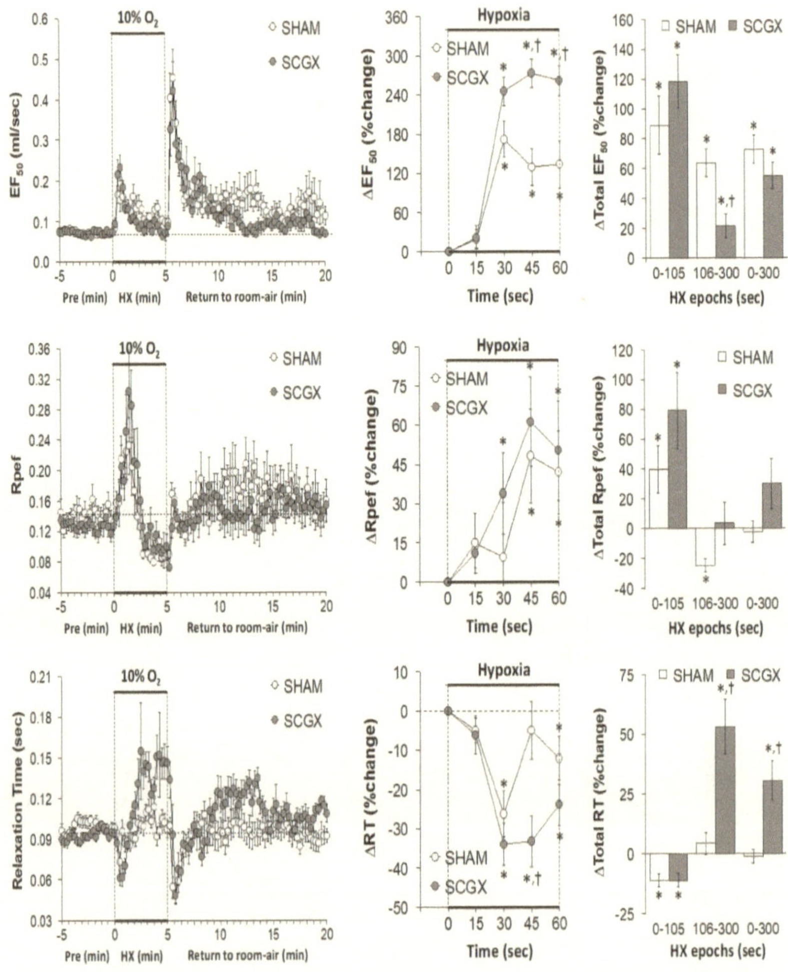

Figure 30. Left panels: EF_{50}, Rpef and relaxation time (RT) values recorded prior to, during and following a hypoxic (HX, 10% O_2, 90% N_2) gas challenge of 5 min in duration in mice that were sham-operated (SHAM) and those with bilateral superior cervical ganglionectomy (SCGX). **Middle Panels:** Responses recorded during the first minute of HXC in SHAM and SCGX rats (expressed as % of pre-values). **Right panels:** Total responses recorded during the first 105 sec, between 106-300 sec and between 0-300 sec (expressed as the sum of all % changes from pre). All results are presented as mean ± SEM. *P < 0.05, significant change from pre-values. †P < 0.05, SCGX *versus* SHAM. There were 14 mice in each of the SHAM and SCGX groups.

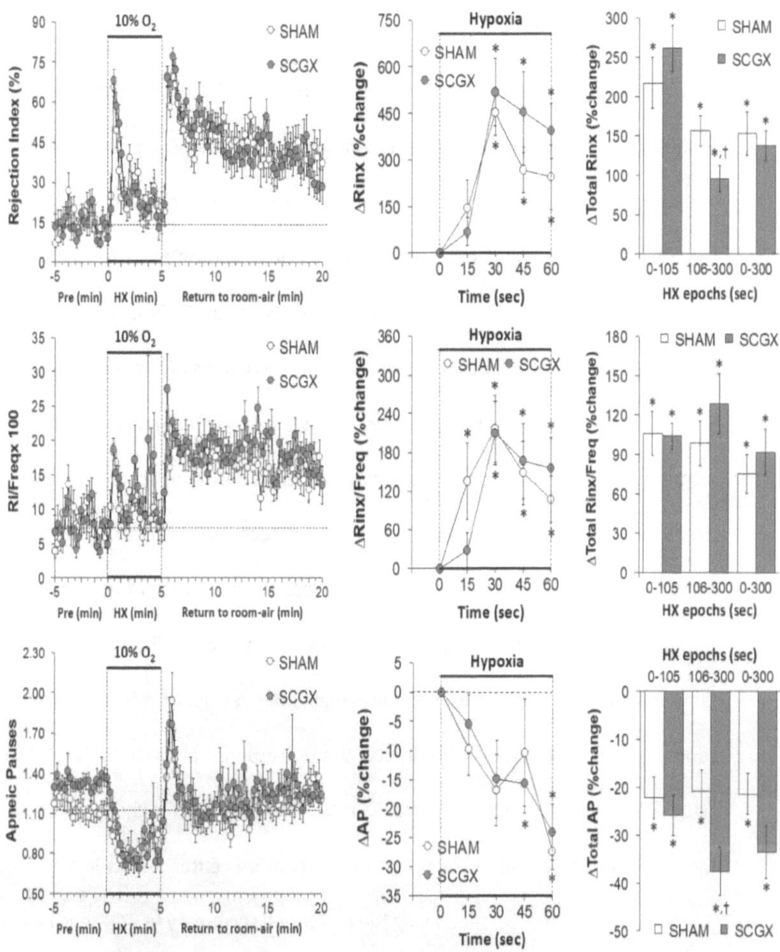

Figure 31. Left panels: Rejection Index (RI, Rinx) values, RI/frequency ratios and Apneic Pause (AP) values recorded prior to, during and following a hypoxic (HX, 10% O_2, 90% N_2) gas challenge of 5 min in duration in mice that were sham-operated (SHAM) and those with bilateral superior cervical ganglionectomy (SCGX). **Middle Panels:** Responses recorded during the first minute of HXC in SHAM and SCGX rats (expressed as % of pre-values). **Right panels:** Total responses recorded during the first 105 sec, between 106-300 sec and between 0-300 sec (expressed as the sum of all % changes from pre). All results are presented as mean ± SEM. *P < 0.05, significant change from pre-values. †P < 0.05, SCGX *versus* SHAM. There were 14 mice in each of the SHAM and SCGX groups.

CHAPTER V: CONCLUSION

In summary this book project describes the role of CSN input on the HX, HC, and HH ventilatory responses in juvenile rats, and also describes how pre- and postganglionic sympathetic innervation of respiratory-related structures, such as the carotid bodies and central chemoreceptors, modulates the ventilatory response to HX in adult mice.

In our first set of novel experiments (Getsy et al., 2020), we found that loss of chemosensory afferent input to respiratory control regions of the brainstem significantly blunts the HX, HC and HH ventilatory responses in juvenile (P25) rats. In addition, our data suggests that prior CSNX does not abolish the functionality of secondary low O_2 sensing structures (i.e., peripheral structures that detect decreases in arterial pO_2 other than the carotid bodies) in the P25 rat that also possess the capability of activating respiratory centers in the brainstem to trigger downstream ventilatory reflexes, however not to the same magnitude as the carotid bodies. Finally, our results of this first study show that additivity of the HX response alone plus the HC response alone (i.e., HX+HC) to give the actual HH response is absent in juvenile rats with CSNX, thus suggesting that loss of CSN chemoafferent signaling to the brainstem respiratory center results in activated central CO_2 chemoreceptors overshadowing the peripheral secondary low O_2 sensing input at the points of convergence between peripheral and central chemoreceptor signals.

Our next experiments elucidated how loss of preganglionic and/or postganglionic fibers in the CSC-SCG complex that innervate respiratory-related structures, such as the carotid bodies and brainstem, affects resting ventilatory parameters and the ventilatory responses to hypoxic gas exposure (Getsy et al., 2021 and 2021). We found that under baseline (normoxic) conditions, the loss of preganglionic sympathetic fibers of the CSC (i.e., CSCX) in adult C57BL6 mice, and the loss of SCG postganglionic sympathetic input (i.e., SCGX), to respiratory-related structures, including the carotid bodies, do not have a noticeable effect on breathing, suggesting that baseline firing of CB glomus cells is not affected by either CSCX or SCGX in such a way that triggers activation of chemoafferents and thus increase in baseline ventilation. Our data also shows that loss of preganglionic sympathetic fibers traveling through the CSC results in an initial increase in ventilatory responses to hypoxic gas challenge in adult C57BL6 mice, therefore suggesting that the CSC may normally provide inhibitory input to peripheral (e.g., carotid body glomus cells) and central (e.g., brainstem) structures involved in the initial expression of the ventilatory response to hypoxia. In addition, we discovered that loss of postganglionic sympathetic fibers from the SCG results in a diminished total ventilatory response to hypoxia in C57BL6 mice compared to the merely increased initial ventilatory response to hypoxia after loss of preganglionic sympathetic fibers, via transection of the CSC.

These findings from our CSC-SCG studies pose multiple questions including, (1) whether hypoxia exposure may directly alter the firing and downstream signaling pattern of SCG cells that are not solely dependent on cholinergic preganglionic input from the CSC, such as the SCG O_2 sensing SIF cells (SIF cells are known to possess (1) afferent connections via the glossopharyngeal nerve (Takaki et al., 2015) that may be adequate to activate the chemoreflex response without CSC activation of SCG postganglionic nerve fibers, and (2) inhibitory dopaminergic synapses on principal SCG postganglionic neurons that may be active during hypoxic stimulation, thus blocking postganglionic sympathetic input to respiratory-related structures, such as the carotid bodies, and blunting the total ventilatory response to hypoxia (Libet and Tosaka, 1970; Tosaka and Kobayashi, 1977)), and (2) which of the principal SCG postganglionic fibers that form synapses with the preganglionic fibers of the CSC are responsible for the neuromodulatory effects of the CSC on the processes that drive ventilatory responses to hypoxia. Ultimately, the novel findings of both our CSCX and SCGX studies and our CSNX study provide support to the idea that loss of either chemosensory or sympathetic input might be involved in the etiology of respiratory disorders, such as sleep-apneas.

CHAPTER VI: FUTURE DIRECTIONS

An important note of this book is that the studies were performed in two different species at varying developmental stages, therefore the responses from the CSNX study could not be compared to the responses of the CSC-SCG studies. It would be very interesting to see if loss of the neuromodulatory function of the CSC-SCG complex can blunt the hypoxic ventilatory response to the same magnitude as loss of CSN chemosensory afferents. The results would possibly suggest a major interplay between the two systems. To rectify this, CSNX studies will need to be perfomed in adult male C57BL6 mice. Nevertheless, we have multiple future directions to explore in the studies presented in this book book.

With regards to the CSNX study (Getsy et al., 2020), it would be interesting to determine whether female rats of age P25 reveal different responses to males at a time when sex hormones are not huge players of physiological status. In addition, it is important to mention that baroreceptor afferents present in the CSN join with those in the aortic depressor nerve to regulate arterial blood pressure (Del Rio et al., 2013), and thus in future studies we need to assess the role of baroreceptor afferents within the CSN and the consequences of their loss (i.e., alterations in cardiovascular responses, sympatho-vagal balance or baroreceptor reflex activity) before and during the HX, HC and HH gas challenges in the SHAM and CSNX rats.

Additionally, there are future experiments that would be logical follow-ups to the CSCX and SCGX studies (Getsy et al., 2021 and 2021). For instance, since the SCG projects to a wide range of peripheral and central structures that control ventilation, we need to better characterize which of the major postganglionic branches of the SCG, namely the ICN, ECN and GGN, regulate the ventilatory responses to hypoxia in the C57BL6 mouse. For instance, bilateral transection of the GGN would help to establish the relevant neuromodulatory effect sympathetic input has on the carotid bodies and carotid sinus in response to hypoxia. Moreover, in order to determine the temporal and genetic aspects of the role of the CSC-SCG complex in the control of ventilation and the response to hypoxia gas challenge, we will repeat the CSCX and SCGX studies of this book at 14 and 30 days in C57BL6 mice, and perform the ECN, ICN and GGN transections at 4, 14 and 30 days in C57BL6 mice, as well as in Swiss-Webster and A/J mice, to determine any genetic differences between strains (Getsy et al., 2014). Finally, we will extend our investigations by evaluating how the expression pattern of various proteins in sympathetic nerve terminals (e.g., tyrosine hydroxylase, fusion proteins mediating vesicular exocytosis, and catecholamine-containing vesicles) and CB glomus cells (e.g., tyrosine hydroxylase, and voltage-gated Na$^+$, K$^+$ and Ca^{2+}-channels) change after transection of the CSC. This set of experiments is especially important since adaptations in protein expression within SCG nerve terminals following CSCX was not determined under resting (i.e., normoxic) and hypoxic conditions in this book study, and may be a major driver behind sympathetic modulation of the chemoafferent response to hypoxia in mammals.

APPENDIX:

MATERIALS AND METHODS

*****All materials and methods were adapted from Getsy et al. 2020, 2021 and 2021.*

PERMISSIONS

All studies were carried out in accordance with the National Institutes of Health "Guide for the Care and Use of Laboratory Animals" (NIH Publication No. 80-23) revised in 1996 (https://www.nap.edu/catalog/5140/guide-for-the-care-and-use-of-laboratory-animals). The protocols were approved by the Institutional Animal Care and Use Committee at Case Western Reserve University.

ANIMALS

<u>CSNX Study</u>

Eighty-two male Sprague Dawley (SD) rats (postnatal age 21, P21) from ENVIGO (Indianapolis, IN, USA) were used. Rats were delivered pathogen free and housed under specific-pathogen free conditions with a 12 h light-dark cycle.

<u>CSCX and SCGX Studies</u>

Adult (12 weeks) C57BL6 male mice were purchased from Jackson Laboratory (Bar Harbor, ME, USA). Mice were delivered pathogen free and housed under specific-pathogen free conditions with a 12h light-dark cycle.

SURGERIES

<u>CSNX Study</u>

P21 SD rats were anesthetized with an intraperitoneal (IP) injection of ketamine (80 mg/kg, Ketaset, Zoetis, Parsippany, NJ) and xylazine (10 mg/kg, Akorn Animal Health, Lake Forest IL), and placed on a surgical station allowing body temperature to be maintained at 37°C via a heating pad (SurgiSuite, Kent Scientific Corporation, Torrington, CT). The surgical plane of anesthesia was checked every 15 min by a toe pinch. A small mid-line incision was made in the neck (approx. 2cm). Dual forceps were used to tease away connective tissue and expose the left and right CSN. The CSN was then cut at the point where it entered the glossopharyngeal nerve (**Figure 32**) by procedures detailed previously (Gaston et al., 2014; Baby et al., 2018). For the SHAM procedure, the left and right CSN were identified, but not transected.

The rats were allowed 4 days to recover from surgery and were P25 on the day of the experimental study. All rats were monitored for pain and distress every day following surgery. Rats were given an injection of the non-steroidal anti-inflammatory drug, carprofen (2 mg/kg, Rimadyl, Zoetis, Parsippany, NJ), 24h and 48h post-surgery to reduce any pain or inflammation at the incision site. None of the rats showed any signs of pain or inflammation from the surgeries and began moving about the cages and eating and drinking approximately 1h after surgery. Rats were weighed daily to ensure proper weight gain. We have determined that these injections of carprofen do not affect resting ventilation or the responses to hypoxia, hypercapnia and hypoxia-hypercapnia gas challenges on day 4 post-surgery (data not shown).

<u>CSCX and SCGX Studies</u>

Mice were anesthetized with an intraperitoneal injection of ketamine (80 mg/kg, Ketaset, Zoetis, Parsippany, NJ) and xylazine (10 mg/kg, Akorn Animal Health, Lake Forest IL), and placed on a surgical station allowing body temperature to be maintained at 37°C via a heating pad (SurgiSuite, Kent Scientific Corporation, Torrington, CT). The adequacy of anesthesia was regularly checked by nociceptive stimulus (toe pinch). A small mid-line incision was made in the neck (approx. 2cm). Dual forceps were used to tease away connective tissue and expose the left and right CSC (**Figure 33**) and SCG (**Figure 34**) behind the carotid artery bifurcation. For the studies involving transection of the left and right CSC, the CSC was exposed and cut using micro-scissors approximately 1 mm from the point where the CSC enters the SCG. For the experiments involving removal of the left and right SCG, the SCG was cut using micro-scissors at the point where the CSC enters the SCG, and the point where the ECN and ICN exit the SCG, and was removed entirely from the animal. In sham-operated (SHAM) mice, for either study, the left and right CSC or SCG were exposed and identified but not cut.

The mice were allowed 4 days to recover from surgery. This time-point for recovery was the same between CSCX and SCGX experiments to allow temporal comparisons between experiments, and was based on substantial evidence that catecholamine levels, identified by tyrosine hydroxylase positive nerve terminals in carotid bodies, are markedly reduced 3-4 days after removal of the ipsilateral SCG (Mir et al., 1982; González-Guerrero et al., 1993; Ichikawa and Helke, 1993; Ichikawa, 2002). All mice were monitored for pain and distress every day following surgery. Mice were given an injection of the non-steroidal anti-inflammatory drug, carprofen (2 mg/kg, Rimadyl, Zoetis, Parsippany, NJ), 24h and 48h post-surgery to reduce any pain or inflammation at the incision site. None of the mice showed any signs of pain or inflammation from the surgeries and began moving about the cages and eating and drinking approximately 1h after surgery. Mice with either bilateral CSCX or SCGX displayed Horner's Syndrome (a condition characterized by constriction of the pupil and caused by damage to sympathetic nerves) after they woke up from the anesthesia. SHAM mice did not display Horner's Syndrome. Mice were weighed daily to ensure proper weight gain. We have determined that these injections of carprofen do not affect resting ventilation or the response to hypoxia gas challenge on day 4 post-surgery (data not shown).

WHOLE BODY PLETHYSMOGRAPHY

Over the years, whole body plethysmography (WBP) has been widely applied to the computation of ventilatory parameters in mice and rats, as an easy and stress-free form of measurement. WBP is an indirect measure of the animal's respiration by monitoring the pressure changes within an unrestrained chamber (**Figure 35a**) which encloses the whole animal. These pressure changes (i.e., box flow waveforms) are a composite of nasal flows and thoracic velocity patterns. WBP then infers and derives the animal's flow parameters from these box flow waveforms (**Figure 35b**). A host of directly recorded parameters and calculated (i.e., derived) parameters from the WBP system include those

listed in **Table 9**. Not all these parameters were used for analysis in each of the

research studies presented in this book.

<u>CSNX Study</u>

Ventilatory parameters were continuously recorded in freely-moving P25 juvenile SHAM

or CSNX rats via whole body plethysmography technology (Buxco® Small Animal Whole

Body Plethysmography, DSI a division of Harvard Biosciences, Inc., St. Paul, MN, USA), as

detailed previously (Baby et al., 2018; Henderson et al., 2013, 2014; May et al., 2013a,b).

All experiments were performed in a quiet laboratory with atmospheric pressure of 760

mmHg (sea-level). The chamber volumes were 0.5 L and the gas flowing through all of the

chambers was 0.5 L/min. Directly recorded parameters chosen from **Table 9** were

frequency of breathing (fR) and tidal volume (VT), and the calculated parameter was

minute ventilation (MV), which is (fR x VT) (Young et al., 2013; Laferrière et al., 2005). The

directly recorded parameters (i.e., fR and VT) were extracted from flow waveforms using

proprietary Biosystem XA and FinePointe software (Data Sciences International, St. Paul,

MN, USA) as described previously (Palmer et al., 2013,a,b; Gaston et al., 2014; Getsy et

al., 2014). All data parameters, both directly recorded or calculated, were extracted as

individual data points (e.g., a fR value for every 15 sec epoch of the recording) from the

FinePointe software and exported into excel spreadsheets. The FinePointe software

constantly corrected digitized values for changes in chamber temperature and humidity,

and a rejection algorithm excluded motion-induced artifacts (Baby et al., 2018;

Henderson et al., 2013, 2014; May et al., 2013a,b; Getsy et al., 2014).

On the day of study, mice were placed in individual whole body plethysmography chambers (Buxco® Small Animal Whole Body Plethysmography, DSI a division of Harvard Biosciences, Inc., St. Paul, MN, USA) to continuously record ventilatory parameters in the freely-moving state, as described in detail previously (Gaston et al., 2014; Getsy et al., 2014; Palmer et al., 2013a,b; Palmer et al., 2015). All studies were performed in a quiet laboratory with atmospheric pressure of 760 mmHg (sea-level). The chamber volumes were 0.5 L and the gas flowing through all of the chambers was 0.5 L/min. The particular array of ventilatory parameters chosen from **Table 9** provide an in-depth analysis of the differences in resting breathing patterns and responses to the hypoxic challenge in SHAM and CSCX mice and SHAM and SCGX mice (Solberg et al., 2006; Moore et al., 2012; Palmer et al., 2013a,b; Gaston et al., 2014; Getsy et al., 2014; Palmer et al., 2015, Villiere et al., 2017). Directly recorded parameters were (a) frequency of breathing (Freq or as in the CSNX study, fR), (b) tidal volume, (TV as in the CSNX study, VT), (c) inspiratory time (Ti) and expiratory time (Te), (d) end inspiratory pause (EIP) and end expiratory pause (EEP), (e) expiratory flow at 50% (EF_{50}), (f) peak inspiratory flow (PIF) and peak expiratory flow (PEF), (g) relaxation time (RT), (h) Rpef, and (i) rejection index (RI). All directly recorded parameters (i.e., Freq, TV, Ti, Te, EIP, EEP, PIF, PEF, Rpef, RT and RI) were extracted from flow waveforms using proprietary Biosystem XA and FinePointe software (Data Sciences International, St. Paul, MN, USA) as described previously (Palmer et al., 2013,a,b; Gaston et al., 2014; Getsy et al., 2014). Calculated parameters included (a) minute ventilation (MV, Freq x TV), (b) Ti/Te and Ti/(Ti+Te), (c) PIF/PEF, (d) inspiratory drive (TV/Ti) and expiratory drive (TV/Te), (e) rejection index corrected for respiratory frequency (RI/Freq), and (f) the number of apneic pauses per epoch, (Te/RT)-1. All data parameters, both directly recorded or calculated, were extracted as individual data points (e.g., a Freq value for every 15 sec epoch of the recording) from the FinePointe software and exported into excel spreadsheets. The FinePointe software constantly corrected digitized values for changes in chamber temperature and humidity, and a rejection algorithm excluded motion-induced artifacts (Baby et al., 2018; Henderson et al., 2013, 2014; May et al., 2013a,b; Getsy et al., 2014).

Because rejection index (RI) was reported in the CSCX and SCGX studies, it is important to explain the meaning behind RI in more detail. The plethysmography system includes a rejection algorithm that was set to reject breaths that did not reflect normal tidal volume breathing, and as such rejected (a) abnormal breaths (abnormal balance of inspiratory and expiratory volumes), apneas, sighs, post-sighs, sniffs, and waveforms that most likely arose from activities such as grooming the face, hands, and rear (Getsy et al., 2014). Minimum TV was set at 0.05 ml, minimum Ti was set at 0.04 sec, maximum Te was set at 0.5 sec, the ratio of inspiratory volume/expiratory volume (volume balance) was set to a range of 90 to 110%. Values were rejected when they fell outside the above criteria and when (a) Ti was greater than 2 times Te, (b) PIF could not be distinguished from PEF, and (c) EF_{50}, Rpef, or RT could not be computed (Getsy et al., 2014).

EXPERIMENTAL PROTOCOLS

<u>CSNX Study – hypoxic, hypercapnic and hypoxic-hypercapnic gas challenge protocols and data collection including maximal attainable responses</u>

Four days post-CSN transection, the P25 rats were placed in plethysmography chambers to continuously record (breath by breath) ventilatory parameters as described above. The rats were allowed to acclimatize for at least 60 min to allow stable baseline values to be recorded over a 15 min period prior to exposing the rats to HX, HC or HH gas challenges. After the rats were placed in the chamber, they would explore the new environment for about 15-20 min after which time they would usually lay still, periodically grooming themselves and occasionally sniffing. As such, at the time the HX (10% O_2, 90% N_2), HC (5% CO_2, 21% O_2, 74% N_2) or HH (5% CO_2, 10% O_2, 85% N_2) gases were delivered to the chambers, the rats were awake and resting quietly. The behavior of the rats did not change appreciably upon delivery of the HX, HC or HH gas challenges. The occasional rat explored for 15-30 sec or groomed for 5-10 sec. The HX, HC, and HH protocols were completed in a controlled fashion. More specifically, pairs of SHAM (n = 2) and CSNX (n = 2) rats were studied on the same day. Because of this experimental paradigm, the collected data are likely to accurately define the effects of CSNX on resting ventilatory parameters and the response to HX, HC or HH gas challenges. It is important to note here that the rats actually underwent five 5-min episodes of a HX, HC or HH gas challenge each separated by 15 min of room-air. We referred to this as 5 x 5 HX, HC or HH gas challenge, however, this book detailed the responses that occurred during the first gas challenge only, with findings pertaining to responses that occurred during challenges 2-5 to be detailed in future analyses. As such, we designated the responses elicited by the first gas challenge as episode 1 (E1). The data was continuously recorded before, during the entire 5 x 5 HX, HC or HH gas challenge, and during each of the five 15-min post-challenge periods. For each ventilatory parameter, data points collected during every 15 sec epoch were averaged for each rat for graphing and analysis purposes. The maximal values during the HX, HC or HH gas challenge did not necessarily occur during the same 15 sec epoch in each rat, and so the maximal values obtained by each rat were also collected. Moreover, the maximal attainable values that occurred throughout the entire study (whether during the 5 x 5 HX, HC or HH gas challenge or subsequent RA phases) were evaluated. Every 15 sec value was sorted from largest to smallest and the top ten highest values were averaged to provide the maximal attainable fR, VT and MV values.

<u>CSCX and SCGX Studies – hypoxic gas challenge protocol and data collection</u>

For either the CSCX study or the SCGX study, SHAM, CSCX and SCGX mice were placed in the plethysmography chambers and allowed approximately 60 min to settle to let resting breathing reach stable levels before exposure to a 5 min HX (10% O_2, 90% N_2) gas challenge and then re-exposure to room-air for 15 min. The behavior of any of the mice did not change appreciably upon delivery of the HX gas challenge. The occasional mouse explored for 15-30 sec or groomed for 5-10 sec. The HX protocol was completed in a controlled fashion. More specifically, pairs of SHAM (n = 2) and CSCX (n = 2) mice or SHAM (n = 2) and SCGX (n = 2) mice were studied on the same day. Because of this experimental paradigm, the collected data are likely to accurately define the effects of CSCX or SCGX on resting ventilatory parameters and the response to HX gas challenge. The data was continuously recorded before, during the entire HX gas challenge, and during the 15 min post-challenge period. For each ventilatory parameter, data points collected during every 15 sec epoch were averaged for each mouse for graphing and analysis purposes.

<u>CSNX Study</u>

All values are expressed as mean ± SEM. The data were analyzed by one-way or two-way analysis of variance (ANOVA) tests that were followed by Student's modified t-tests with Bonferroni corrections for multiple comparisons between means using the error mean square terms arising from the ANOVA tests (Wallenstein et al., 1980) as detailed previously (May et al., 2013a,b). The ANOVAs and the multiple comparisons tests were performed using the statistical programs, JMP Statistical Analysis Software (JMP, Cary, NC, USA) and Sigma XL (SigmaStat, Canada, Kitchener, Ontario, Canada). In every instance, these programs tested for normality and homogeneity of variances before running the ANOVAs. All data sets were normally distributed with homogenous variances and therefore suitable for ANOVA testing. For each rat, the pre-values for each parameter were determined as the average of the values recorded over the 90 seconds (i.e., six 15-sec epochs) immediately prior to each gas challenge. These resting (baseline) data, and the data recorded over the entire acclimatization period, from the six groups of rats (SHAM and CSNX groups for the HX, HC and HH gas challenges) were analyzed by one-way ANOVA with Bonferroni corrections. The arithmetic changes in ventilatory parameters elicited by the gas challenges at 15, 30, 45 and 60 sec were determined by simply subtracting the actual values obtained at these time points for each rat from the pre-value recorded for each rat. These individual values at 15, 30, 45 and 60 sec were summed to provide the total of these responses. The resulting three values for each SHAM and CSNX groups (HX, HC, and HH values) were then analyzed by one-way ANOVA with Bonferroni corrections. The total arithmetic changes in ventilatory parameters in SHAM and CSNX rats were calculated from pre-values and summed to provide the total response (i.e., sum of 20 individual 15 sec epochs for the 5 min challenge). The resulting three values for each of the SHAM and CSNX groups (HX, HC, HH values) were then analyzed by one-way ANOVA with Bonferroni corrections. To compare the ventilatory parameters recorded during the actual HH challenge with the additive HX+HC values, we performed simple addition of responses obtained from the rats that received the HX challenge and from rats that received the HC challenge. These data are presented as mean of summed values ± SEM (10% of mean). The 10% SEM value was chosen because it is consistent with the SEM values of the actual data. It is important to note that none of the theoretical HX+HC values were used in any of the statistical analyses as the HX+HC values were only shown to support the interpretation of the actual HX, HC and HH responses. For descriptive purpose only, we chose a percentage difference of greater or less than 25% of the actual HH values to be a noticeable change from the HX+HC values and therefore subject to discussion.

<u>CSCX and SCGX Studies</u>

A data point before, during and after the HX gas challenge was collected every 15 sec for each ventilatory parameter. The arithmetic changes in ventilatory parameters elicited by the HXC at 15, 30, 45 and 60 sec were determined by simply subtracting the actual values obtained at these time points for each mouse (i.e., each SHAM and CSCX mouse and each SHAM and SCGX mouse) from the pre-value recorded for each mouse. To determine the total responses (cumulative % changes from pre-HXC values) during the hypoxic gas challenge and return to room-air for each mouse, we used the formulas, (a) total hypoxic response = (sum of the 20 values during hypoxic challenge) − (pre-value x 20), and (b) total room-air response = (sum of the 60 values during room-air) (mean of pre-values x 60). The mean and SEM for the SHAM and CSCX mice and SHAM and SCGX groups was calculated. There were 20 values (i.e., data points) for each mouse recorded during the hypoxic challenge (4 15-sec data points per minute x 5-minute recording period = 20 data points). Similarly, 60 values were collected for the post-hypoxic (room-air) phase for each mouse (4 15-sec data points per minute x 15-minute recording period = 60 data points). We also determined the total responses during the HXC 0-105 sec epoch and 106-300 sec epoch, in addition to the entire HXC 5 min (0-300 sec epoch). Note that during the HXC, the epoch 106-300 sec in which the demonstrated differences between SHAM and SCGX mice were most evident, was associated with normal (pre-HXC) levels of rejection index in both groups (SHAM and CSCX and SHAM and SCGX) and as such the differences between the groups were not complicated by any data sampling issues. All data are presented as mean ± SEM. All data were analyzed by one-way or two-way ANOVA followed by Student's modified t-test with Bonferroni corrections for multiple comparisons between means (Palmer et al., 2013a,b).

Table 9: Directly recorded ventilatory parameters except those (***) derived from formulae

Parameter	Abbreviation	Units	Definition
Directly recorded parameters			
1. Frequency of breaths	Freq	breaths/min	Rate of breathing
2. Inspiratory Time	T_I	sec	Duration of inspiration
3. Expiratory Time	T_E	sec	Duration of expiration
4. End Inspiratory Pause	EIP	msec	Pause between end of inspiration and start of expiration
5. End Expiratory Pause	EEP	msec	Pause between end of expiration and start of inspiration
6. Relaxation time	RT	sec	Decay of expiration to 36% maximum
7. Rpef	Rpef = (Time to PEF)/Te	No units	Relative rate of achieving peak expiratory flow
8. Tidal Volume	TV	ml	Volume of inspired air per breath
9. Minute Ventilation	MV = freq x TV	ml/min	Total volume of air inspired per min
10. Peak Inspiratory Flow	PIF	ml/sec	Maximum inspiratory flow
11. Peak Expiratory Flow	PEF	ml/sec	Maximum expiratory flow
12. Expiratory flow at 50%	EF_{50}	mls/sec	Expiratory flow at 50% expired tidal volume
13. Chamber Temperature	Temp	°C	Direct measure of chamber temperature
14. Chamber Humidity	Hum	%	Direct measure of chamber humidity
15. Rejection Index	RI	%	Percentage of non-eupneic breaths per epoch
*****Derived parameters**			
16. Inspiratory Drive	TV/Ti	ml/sec	Central urge to inhale
17. Expiratory Drive	TV/Te	ml/sec	Central drive to exhale
18. Ti/Te	Ti/Te	No units	Ratio of inspiratory to expiratory time
19. Ti/(Ti + Te)	Ti/(Ti + Te)	no units	Inspiratory quotient
20. PIF/PEF	PIF/PEF	No units	Ventilatory balance quotient
21. RI/Frequency	RI/Freq	%/(breaths/min	Balanced rejection index
22. Apneic Pause	AP = (Te/RT)-1	No units	No of apneic pauses

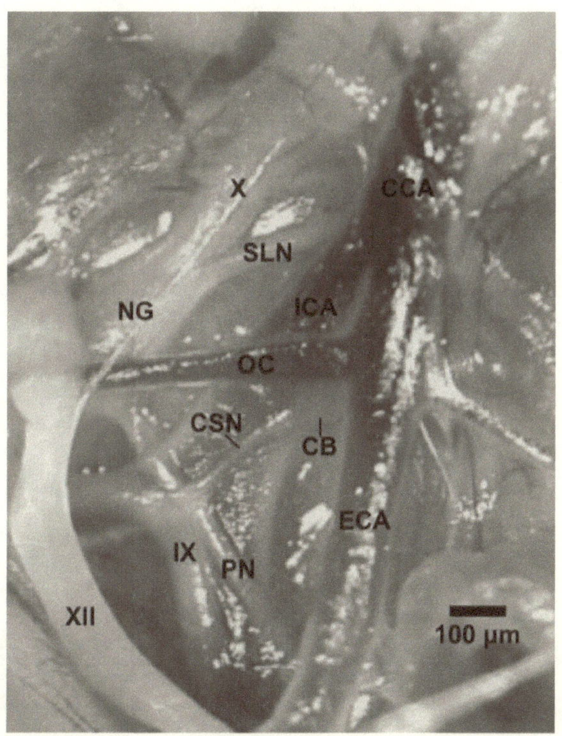

Figure 32. Photograph in a SD male P25 juvenile rat of the carotid sinus nerve (CSN) branching off the glossopharyngeal nerve (IX) and entering the carotid body (CB). The pharyngeal nerve (PN), hypoglossal nerve (XII), nodose ganglion (NG), vagus nerve (X), superior laryngeal nerve (SLN), common carotid artery (CCA), internal (ICA) and external (ECA) carotid arteries, and occipital artery (OC) are also shown. This dissection was done on the left side of the animal. The scale bar is 100 m.`

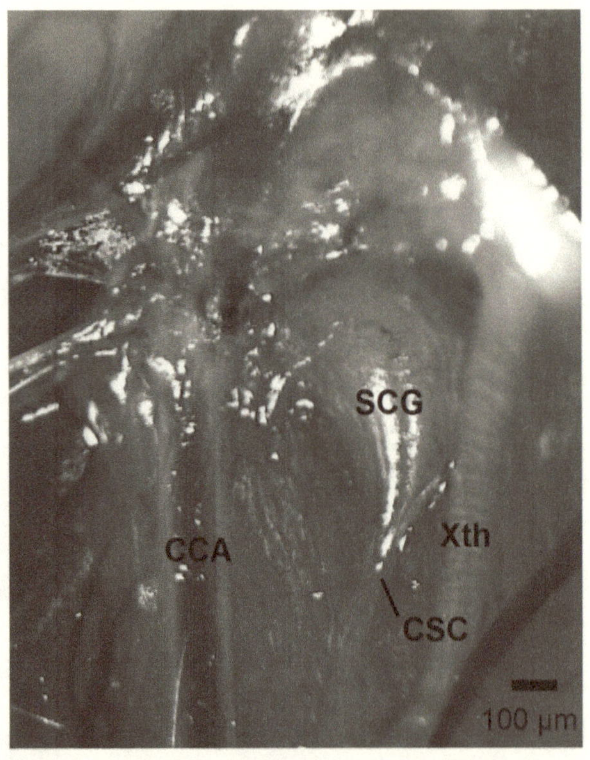

Figure 33. Photograph of the left cervical sympathetic chain (CSC) left superior cervical ganglion (SCG) and left internal (ICN) and external (ECN) nerves emanating from the SCG of a C57BL6 mouse. The vagus nerve (X) and common carotid artery (CCA) are also shown.
The scale bar is 100 m.`

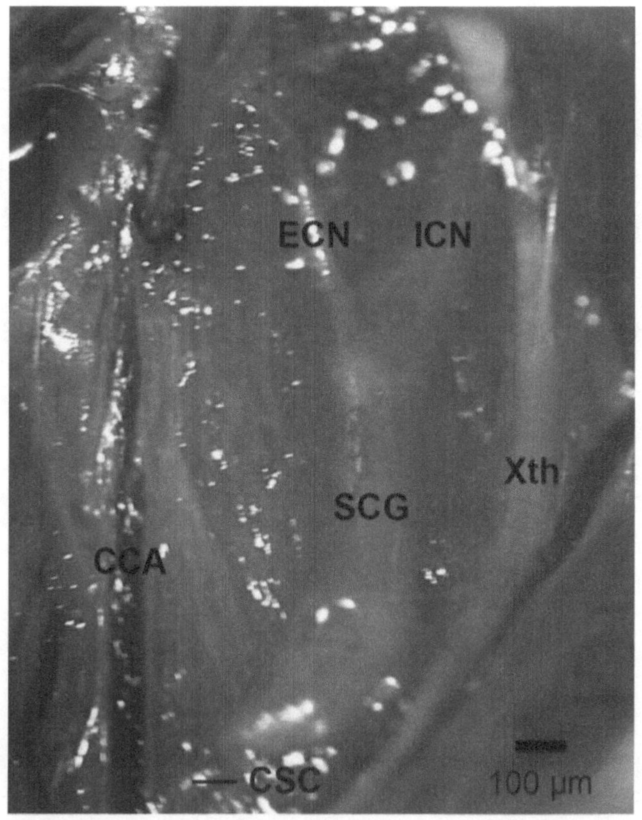

Figure 34. Picture of the left superior cervical ganglion (SCG) of a C57BL6 mouse and the associated cervical sympathetic chain (CSC), left internal carotid nerve (ICN) and external carotid nerve (ECN) emanating from the SCG. The common carotid artery (CCA) and vagus (Xth cranial) nerve and are shown. The scale bar in the photograph represents a length of 100 µm.

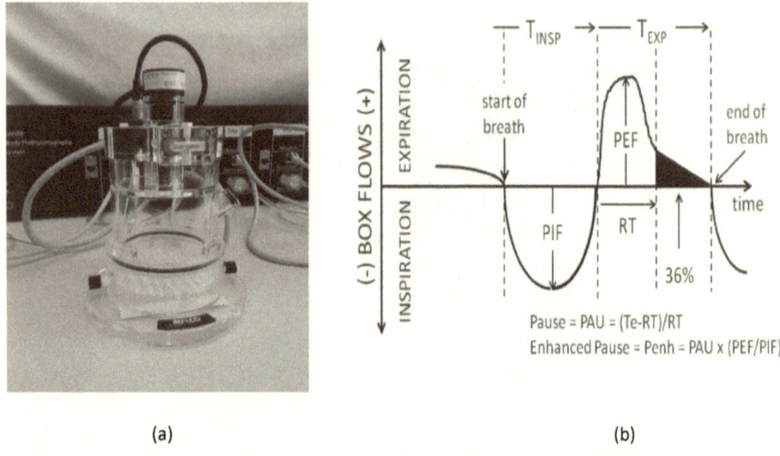

(a)   (b)

Figure 35: Whole body plethysmography system set-up and schematic of breathing waveform. (a) Image of WBP chamber taken in Dr. Stephen J Lewis' Lab at Case Western Reserve University. (b) Schematic flow waveform from WBP system. Abbreviations: T_{INSP}, Inspiratory Time; T_{EXP}, Expiratory Time; PIF, Peak Inspiratory Flow; PEF, Peak Expiratory Flow; RT, Relaxation Time.

Date: **December 15, 2020**

Name: **Stephen J Lewis**

Title: **Breathing pattern variability in response to hypoxic and hypercapnic gas challenges in rats**

Protocol Number: **2015-0005**

Species: **Rat 8,925**

This Institution's Animal Care and Use Committee reviewed and approved the Animal Experimentation 2015-0005. TheCommittee has approved this protocol for a period of three years, beginning **12/15/2020** through and ending **12/15/2023**.

Before ordering animals, it is important to confirm that all animal handlers have completed their necessary training and have access to the animal facility. If animals are ordered prematurely, your staff will not have access to the animals or be able to schedule procedures associated with this protocol. For information on Training Requirements and Facility Access please visit theIACUC Training page https://case.edu/medicine/research/research-support/iacuc

While the approved protocol covers 3 years, annual updates are due on the anniversary of the original approval. You will receivea continuing review email notification from eSirius3G sixty days prior to your annual due date and an Alert will remain in your eSirius3G Dashboard Mailbox until completed. You can access, complete, and submit your Continuing Review form directly from your eSirius3G Dashboard.

Please remember that any change to the scope of your animal protocol must be submitted to the IACUC office for Committee review and approval. No work may begin on a proposed addendum until the principal investigator has received either an emailnotification of approval through eSirius3G or a final written approval letter signed by the IACUC chair.

If this protocol supports a federally funded grant or sponsored project you must alert your office of grants and contracts if this protocol expires or is otherwise terminated. You must alert grants and contracts if a new protocol is approved to replace or add tothe current protocol.
To aid in extramural grant submissions, the following description may be used as a guide for NIH and other grant submissions:

The animals described in this study will be housed in the AAALAC accredited facilities of the CWRU School of Medicine (SOM). Standard Operating Procedures and reference materials are available from the IACUC Office for animal use. The animalhealth program for all Case owned laboratory animals is directed by the Case Animal Resource Center Director, W. John Durfee,DVM, Diplomate ACLAM, and provided by two full-time clinical veterinarians. Animals in each room are observed daily for signs of illness by the animal technician responsible for providing husbandry. Medical records and documentation of experimental use are maintained individually for non-rodents and individually or by cage group for rodents. Veterinary technicians under the direction of the attending veterinarian provide routine veterinary medical care to all animals. Animal care and use is additionally monitored for training and compliance issues by the Director, Research Compliance IACUC. The Case Assurance number is A-3145-01, valid until April 30, 2023.

Sincerely yours,

Efstathios
Karathana
sisIACUC
Chair

REFERENCES:

Alberola-Die, A., Reboreda, A., Lamas, J. A., and Morales, A. (2013). Lidocaine effects on acetylcholine-elicited currents from mouse superior cervical ganglion neurons. *Neurosci. Res.* 75, 198-203. doi: 10.1016/j.neures.2013.01.005

Alcayaga, J., Cerpa, V., Retamal, M., Arroyo, J., Iturriaga, R., and Zapata, P. (2000). Adenosine triphosphate-induced peripheral nerve discharges generated from the cat petrosal ganglion in vitro. *Neurosci Lett.* 282, 185–188. doi: 10.1016/s0304-3940(00)00896-x

Almaraz, L., Pérez-García, M. T., Gómez-Nino, A., and González, C. (1997). Mechanisms of alpha$_2$-adrenoceptor-mediated inhibition in rabbit carotid body. *Am. J. Physiol.* 272, C628-C637. doi: 10.1152/ajpcell.1997.272.2.C628

Annese, V., Navarro-Guerrero, E., Rodríguez-Prieto, I., and Pardal, R. (2017). Physiological Plasticity of Neural-Crest-Derived Stem Cells in the Adult Mammalian Carotid Body. *Cell Rep.* 19, 471–478. doi: 10.1016/j.celrep.2017.03.065

Asamoto, K. (2004). Neural circuit of the cervical sympathetic nervous system with special reference to input and output of the cervical sympathetic ganglia: relationship between spinal cord and cervical sympathetic ganglia and that between cervical sympathetic ganglia and their target organs. *Kaibogaku Zasshi.* 79, 5-14.

Atanasova, D. Y., Iliev, M. E., Nikolai, E., and Lazarov, N. E. (2011). Morphology of the rat carotid body. *Biomed Rev.* 22, 41–55. doi: 10.14748/bmr.v22.34

Baby, S. M., Gruber, R. B., Young, A. P., MacFarlane, P. M., Teppema, L. J., and Lewis, S. J. (2018). Bilateral carotid sinus nerve transection exacerbates morphine-induced respiratory depression. *Eur. J. Pharmacol*. 834, 17-29. doi:

10.1016/j.ejphar.2018.07.018

Banks, B. E., and Walter, S. J. (1975). Proceedings: The effects of axotomy and Nerve

Growth Factor on the neuronal population of the superior cervical ganglion of the

mouse. *J. Physiol*. 249, 61P-62P.

Bascom, A. T., Sankari, A., and Badr, M. S. (2016). Spinal cord injury is associated with

enhanced peripheral chemoreflex sensitivity. *Physiol. Rep*. 4, e12948. doi:

10.14814/phy2.12948

Benot, A. R., and López-Barneo, J. (1990). Feedback Inhibition of Ca^{2+} Currents by

Dopamine in Glomus Cells of the Carotid Body. *Eur J Neurosci*. 2, 809–812. doi:

10.1111/j.1460-9568.1990.tb00473.x

Berlowitz, D. J., Wadsworth, B., and Ross, J. (2016). Respiratory problems and

management in people with spinal cord injury. *Breathe (Sheff)*. 12, 328-340. doi:

10.1183/20734735.012616

Bernardini, A., Wolf, A., Brockmeier, U., Riffkin, H., Metzen, E., Acker-Palmer, A.,

Fandrey, J., and Acker, H. (2020). Carotid body type I cells engage flavoprotein and

Pin1 for oxygen sensing. *Am. J. Physiol. Cell Physiol*. 318, C719-C731. doi:

10.1152/ajpcell.00320.2019

Biscoe, T. J., and Purves, M. J. (1967). Observations on carotid body chemoreceptor

activity and cervical sympathetic discharge in the cat. *J. Physiol*. 190, 413-424. doi:

10.1113/jphysiol.1967.sp008218

Bisgard, G. E., Mitchell, R. A., and Herbert, D. A. (1979). Effects of dopamine, norepinephrine and 5-hydroxytryptamine on the carotid body of the dog. *Respir. Physiol.* 37, 61-80. doi: 10.1016/0034-5687(79)90092-6

Bisgard, G., Warner, M., Pizarro, J., Niu, W., and Mitchell, G. (1993). Noradrenergic inhibition of the goat carotid body. *Adv. Exp. Med. Biol.* 337, 259-263. doi: 10.1007/978-1-4615-2966-8_36

Bisgard, G. E., Forster, H. V., Orr, J. A., Buss, D. D., Rawlings, C. A., and Rasmussen, B. (1976). Hypoventilation in ponies after carotid body denervation. *J. Appl. Physiol.* 40, 184–190. doi: 10.1152/jappl.1976.40.2.184

Bisgard, G. E., Forster, H. V., and Klein, J. P. (1980). Recovery of peripheral chemoreceptor function after denervation in ponies. *J. Appl. Physiol. Respir. Environ. Exerc. Physiol.* 49, 964–970. doi: 10.1152/jappl.1980.49.6.964

Bisgard G. E. (2000). Carotid body mechanisms in acclimatization to hypoxia. *Respir Physiol.* 121, 237–246. doi: org/10.1016/s0034-5687(00)00131-6

Black, I. B., Hendry, I. A., and Iversen, L. L. (1972). The role of post-synaptic neurones in the biochemical maturation of presynaptic cholinergic nerve terminals in a mouse sympathetic ganglion. *J. Physiol.* 221, 149-159. doi: 10.1113/jphysiol.1972.sp009745

Blain, G. M., Smith, C. A., Henderson, K. S., and Dempsey, J. A. (2010). Peripheral chemoreceptors determine the respiratory sensitivity of central chemoreceptors to CO_2. *J. Physiol.* 588, 2455-2471. doi: 10.1113/jphysiol.2010.187211

Bollé, T., Lauweryns, J. M., and Lommel, A. V. (2000). Postnatal maturation of neuroepithelial bodies and carotid body innervation: a quantitative investigation in the rabbit. *J Neurocytol.* 29, 241–248. doi: 10.1023/a:1026567603514

Bolter, C. P., and Ledsome, J. R. (1976). Effect of cervical sympathetic nerve stimulation on canine carotid sinus reflex. *Am. J. Physiol.* 230, 1026-1030. doi: 10.1152/ajplegacy.1976.230.4.1026

Bonora, M., Marlot, D., Gautier, H., and Duron, B. (1984). Effects of hypoxia on ventilation during postnatal development in conscious kittens. *J. Appl. Physiol. Respir. Environ. Exerc. Physiol.* 56, 1464–1471. doi: 10.1152/jappl.1984.56.6.1464

Borghini, N., Dalmaz, Y., and Peyrin, L. (1991). Effect of guanethidine on dopamine in small intensely fluorescent cells of the superior cervical ganglion of the rat. *J. Auton. Nerv. Syst.* 32, 13-19. doi: 10.1016/0165-1838(91)90230-z

Boron, W. F., and Boulpaep, E. L. *Medical Physiology: A Cellular and Molecular Approach*. Philadelphia, PA., Saunders Elsevier, 2012.

Bowers, C. W., and Zigmond, R. E. (1979). Localization of neurons in the rat superior cervical ganglion that project into different postganglionic trunks. *J. Comp. Neurol.* 185, 381-391. doi: 10.1002/cne.901850211

Brognara F., Felippe I.S.A., Salgado H.C., and Paton J.F.R. (2020). Autonomic Innervation of the Carotid Body as a Determinant of its Sensitivity - Implications for Cardiovascular Physiology and Pathology. *Cardiovasc. Res.* 2020 Aug 24:cvaa250. doi: 10.1093/cvr/cvaa250

Brokaw, J. J., and Hansen, J. T. (1987). Evidence that dopamine regulates norepinephrine synthesis in the rat superior cervical ganglion during hypoxic stress. *J. Auton. Nerv. Syst.* 18, 185-193. doi: 10.1016/0165-1838(87)90117-2

Brown, D. R., Forster, H. V., Greene, A. S., and Lowry, T. F. (1993). Breathing periodicity in intact and carotid body-denervated ponies during normoxia and chronic hypoxia. *J. Appl. Physiol. (1985)*. 74, 1073-1082. doi: 10.1152/jappl.1993.74.3.1073

Buckler, K. J., Williams, B. A., and Honore, E. (2000). An oxygen-, acid- and anaesthetic-sensitive TASK-like background potassium channel in rat arterial chemoreceptor cells. *J Physiol*. 525, 135–142. doi: 10.1111/j.1469-7793.2000.00135.x

Buckler, K. J., and Vaughan-Jones, R. D. (1994). Effects of hypercapnia on membrane potential and intracellular calcium in rat carotid body type I cells. *J Physiol*. 478, 157–171. doi: 10.1113/jphysiol.1994.sp020239

Buckler, K. J., and Turner, P. J. (2013). Oxygen sensitivity of mitochondrial function in rat arterial chemoreceptor cells. *J Physiol*. 591, 3549-3563. doi: 10.1113/jphysiol.2013.257741

Buller, K. M., and Bolter, C. P. (1993). The localization of sympathetic and vagal neurones innervating the carotid sinus in the rabbit. *J. Auton. Nerv. Syst.* 44, 225-231. doi: 10.1016/0165-1838(93)90035-s

Buller, K. M., and Bolter, C. P. (1997). Carotid bifurcation pressure modulation of spontaneous activity in external and internal carotid nerves can occur in the superior cervical ganglion. *J. Auton. Nerv. Syst.* 67, 24-30. doi: 10.1016/s0165-1838(97)00088-x

Buttigieg, J., and Nurse, C. A. (2004). Detection of hypoxia-evoked ATP release from chemoreceptor cells of the rat carotid body. *Biochem Biophys Res Commun*. 322, 82–87. https://doi.org/10.1016/j.bbrc.2004.07.081

Bureau, M. A., Zinman, R., Foulon, P., and Begin, R. (1984). Diphasic ventilatory response
to hypoxia in newborn lambs. *J. Appl. Physiol. Respir. Environ. Exerc. Physiol*. 56, 84–
90. doi: 10.1152/jappl.1984.56.1.84

Bureau, M. A., Lamarche, J., Foulon, P., and Dalle, D. (1985). The ventilatory response to
hypoxia in the newborn lamb after carotid body denervation. *Respir. Physiol*. 60,
109–119. doi: 10.1016/0034-5687(85)90043-x

Cadaveira-Mosquera, A., Pérez, M., Reboreda, A., Rivas-Ramírez, P., Fernández-
Fernández, D., and Lamas, J. A. (2012). Expression of K2P channels in sensory and
motor neurons of the autonomic nervous system. *J. Mol. Neurosci*. 48, 86-96. doi:
10.1007/s12031-012-9780-y

Campanucci, V. A., and Nurse, C. A. (2007). Autonomic innervation of the carotid body:
role in efferent inhibition. *Respir Physiol Neurobiol*. 157, 83–92. doi:
10.1016/j.resp.2007.01.020

Campen, M. J., Tagaito, Y., Li, J., Balbir, A., Tankersley, C. G., Smith, P., Schwartz, A., and
O'Donnell, C. P. (2004). Phenotypic variation in cardiovascular responses to acute
hypoxic and hypercapnic exposure in mice. *Physiol. Genomics*. 20, 15-20. doi:
10.1152/physiolgenomics.00197.2003

Campen, M. J., Tagaito, Y., Jenkins, T. P., Balbir, A., and O'Donnell, C. P. (2005). Heart
rate variability responses to hypoxic and hypercapnic exposures in different mouse
strains. *J. Appl. Physiol. (1985)*. 99, 807-813. doi: 10.1152/japplphysiol.00039.2005

Cardenas, H., and Zapata, P. (1983). Ventilatory reflexes originated from carotid and extracarotid chemoreceptors in rats. Am. J. Physiol. 244, R119–R125. doi: 10.1152/ajpregu.1983.244.1.R119

Cardinali, D. P., Vacas, M. I., and Gejman, P. V. (1981a). The sympathetic superior cervical ganglia as peripheral neuroendocrine centers. *J. Neural. Transm.* 52, 1-21. doi: 10.1007/BF01253092

Cardinali, D. P., Vacas, M. I., Luchelli de Fortis, A., and Stefano, F. J. (1981b). Superior cervical ganglionectomy depresses norepinephrine uptake, increases the density of alpha-adrenoceptor sites, and induces supersensitivity to adrenergic drugs in rat medial basal hypothalamus. *Neuroendocrinology.* 33, 199-206. doi: 10.1159/000123229

Cardinali, D. P., Pisarev, M. A., Barontini, M., Juvenal, G. J., Boado, R. J., and Vacas, M. I. (1982). Efferent neuroendocrine pathways of sympathetic superior cervical ganglia. Early depression of the pituitary-thyroid axis after ganglionectomy. *Neuroendocrinology.* 35, 248-254. doi: 10.1159/000123390

Chahine, R., Nadeau, R., Lamontagne, D., Yamaguchi, N., and de Champlain, J. (1994). Norepinephrine and dihydroxyphenylglycol effluxes from sympathetic nerve endings during hypoxia and reoxygenation in the isolated rat heart. *Can. J. Physiol. Pharmacol.* 72, 595-601. doi: 10.1139/y94-085

Chai, S., Gillombardo, C. B., Donovan, L., and Strohl, K. P. (2011). Morphological differences of the carotid body among C57/BL6 (B6), A/J, and CSS B6A1 mouse strains. *Respir. Physiol. Neurobiol.* 177, 265–272. doi: 10.1016/j.resp.2011.04.021

Chalmers, J. P., Korner, P. I., and White, S. W. (1967). The relative roles of the aortic and

carotid sinus nerves in the rabbit in the control of respiration and circulation during

arterial hypoxia and hypercapnia. *J. Physiol.* 188, 435–450. doi:

10.1113/jphysiol.1967.sp008148

Coles, S. K., and Dick, T. E. (1996). Neurones in the ventrolateral pons are required for

post-hypoxic frequency decline in rats. *J. Physiol.* 497, 79–94. doi:

10.1113/jphysiol.1996.sp021751

Collins, H. L., Rodenbaugh, D. W., and DiCarlo, S. E. (2006). Spinal cord injury alters

cardiac electrophysiology and increases the susceptibility to ventricular

arrhythmias. *Prog. Brain Res.* 152, 275–288. doi: 10.1016/S0079-6123(05)52018-1

Cummings, K. J., Li, A., Deneris, E. S., and Nattie, E. E. (2010). Bradycardia in serotonin-

deficient Pet-1-/- mice: influence of respiratory dysfunction and hyperthermia over

the first 2 postnatal weeks. *Am J Physiol Regul Integr Comp Physiol.* 298, R1333–

R1342. doi: 10.1152/ajpregu.00110.2010

Dalmaz, Y., Borghini, N., Pequignot, J.M., and Peyrin, L. (1993). Presence of

chemosensitive SIF cells in the rat sympathetic ganglia: a biochemical,

immunocytochemical and pharmacological study. *Adv. Exp. Med. Biol.* 337, 393-

399. doi: 10.1007/978-1-4615-2966-8_55

David, R., Ciuraszkiewicz, A., Simeone, X., Orr-Urtreger, A., Papke, R. L., McIntosh, J. M.,

Huck, S., and Scholze, P. (2010). Biochemical and functional properties of distinct

nicotinic acetylcholine receptors in the superior cervical ganglion of mice with

targeted deletions of nAChR subunit genes. *Eur. J. Neurosci.* 31, 978-993. doi:

10.1111/j.1460-9568.2010.07133.x

Dart, A. M., and Riemersma, R. A. (1991). Noradrenaline release from the rat heart during anoxia: effects of changes in extracellular sodium concentration and inhibition of sodium uptake mechanisms. *Clin. Exp. Pharmacol. Physiol.* 18, 43-46. doi: 10.1111/j.1440-1681.1991.tb01375.x

Del Rio, R., Marcus, N. J., and Schultz, H. D. (2013). Carotid chemoreceptor ablation improves survival in heart failure: rescuing autonomic control of cardiorespiratory function. *J Am Coll Cardiol.* 62, 2422–2430. doi: 10.1016/j.jacc.2013.07.079

Del Rio, R., Andrade, D.C., Lucero, C., Arias, P., and Iturriaga, R. (2016) Carotid Body Ablation Abrogates Hypertension and Autonomic Alterations Induced by Intermittent Hypoxia in Rats. *Hypertension.* 68, 436-445. doi: 10.1161/HYPERTENSIONAHA.116.07255

Dempsey, J. A. (2005). Crossing the apnoeic threshold: causes and consequences. *Exp. Physiol.* 90, 13–24. doi: 10.1113/expphysiol.2004.028985

Deng, B. S., Nakamura, A., Zhang, W., Yanagisawa, M., Fukuda, Y., and Kuwaki, T. (2007). Contribution of orexin in hypercapnic chemoreflex: evidence from genetic and pharmacological disruption and supplementation studies in mice. *J Appl Physiol (1985).* 103, 1772–1779. doi: 10.1152/japplphysiol.00075.2007

Dick, T. E., and Coles, S. K. (2000). Ventrolateral pons mediates short-term depression of respiratory frequency after brief hypoxia. *Respir. Physiol.* 121, 87–100. doi: 10.1016/s0034-5687(00)00121-3

Dillon, G. H., and Waldrop, T. G. (1993). Responses of feline caudal hypothalamic cardiorespiratory neurons to hypoxia and hypercapnia. *Exp. Brain Res*. 96, 260-272. doi: 10.1007/BF00227106

DiMarco, A. F., Kowalski, K. E., Geertman, R. T., Hromyak, D. R., Frost, F. S., Creasey, G. H., & Nemunaitis, G. A. (2009). Lower thoracic spinal cord stimulation to restore cough in patients with spinal cord injury: results of a National Institutes of Health-Sponsored clinical trial. Part II: clinical outcomes. *Arch. Phys. Med. Rehabil*. 90, 726–732. doi: 10.1016/j.apmr.2008.11.014

Dinger, B., Wang, Z. Z., Chen, J., Wang, W. J., Hanson, G., Stensaas, L. J., and Fidone, S. J. (1993). Immunocytochemical and neurochemical aspects of sympathetic ganglion chemosensitivity. *Adv. Exp. Med. Biol*. 337, 25–30

Du, X. J., Bobik, A., Esler, M. D., and Dart, A. M. (1997). Effects of intracellular Ca2+ chelating on noradrenaline release in normoxic and anoxic hearts. *Clin. Exp. Pharmacol. Physiol*. 24, 819–823. doi: 10.1111/j.1440-1681.1997.tb02697.x

Dumas, T. C. (2005). Late postnatal maturation of excitatory synaptic transmission permits adult-like expression of hippocampal-dependent behaviors. *Hippocampus*. 15, 562-578. doi: 10.1002/hipo.20077

Easton, P. A., Slykerman, L. J., and Anthonisen, N. R. (1986). Ventilatory response to sustained hypoxia in normal adults. *J. Appl. Physiol*. (1985). 61, 906–911. doi: 10.1152/jappl.1986.61.3.906

Eccles, R. M. and Libet, B. (1961). Origin and blockade of the synaptic responses of curarized sympathetic ganglia. *J. Physiol*. 157, 484-503. doi: 10.1113/jphysiol.1961.sp006738

Eden, G. J., and Hanson, M. A. (1987). Maturation of the respiratory response to acute hypoxia in the newborn rat. *J. Physiol.* 392, 1–9. doi: org/10.1113/jphysiol.1987.sp016765

Eldridge, F. L., and Gill-Kumar, P. (1980). Mechanisms of hyperpnea induced by isoproterenol. *Respir. Physiol.* 40, 349-363. doi: 10.1016/0034-5687(80)90034-1

Eldridge, F. L. (1974). Central neural respiratory stimulatory effect of active respiration. *J. Appl. Physiol.* 37, 723–735. doi: 10.1152/jappl.1974.37.5.723

Eldridge, F. L., Gill-Kumar, P., and Millhorn, D. E. (1981). Input-output relationships of central neural circuits involved in respiration in cats. *J. Physiol.* 311, 81-95. doi: 10.1113/jphysiol.1981.sp013574

El-Fadaly, A. B., and Kummer, W. (2003). The spatial relationship between type I glomus cells and arteriolar myocytes in the mouse carotid body. *Ann. Anat.* 185, 507–515. doi: 10.1016/S0940-9602(03)80114-0

Elnazir, B., Marshall, J. M., and Kumar, P. (1996). Postnatal development of the pattern of respiratory and cardiovascular response to systemic hypoxia in the piglet: the roles of adenosine. *J. Physiol.* 492, 573–585. doi: 10.1113/jphysiol.1996.sp021330

Erickson, J. T., Mayer, C., Jawa, A., Ling, L., Olson, E. B., Jr, Vidruk, E. H., Mitchell, G. S., and Katz, D. M. (1998). Chemoafferent degeneration and carotid body hypoplasia following chronic hyperoxia in newborn rats. *J Physiol.* 509, 519–526. doi: 10.1111/j.1469-7793.1998.519bn.x

Esquifino, A. I., Alvarez, M. P., Cano, P., Jiménez, V., and Duvilanski, B. (2004). Superior cervical ganglionectomy differentially modifies median eminence and anterior and

mediobasal hypothalamic GABA content in male rats: effects of
hyperprolactinemia. *Exp. Brain. Res.* 157, 296–302. doi: 10.1007/s00221-004-1843-z

EYZAGUIRRE, C., and LEWIN, J. (1961). The effect of sympathetic stimulation on carotid
nerve activity. *J Physiol.* 159, 251–267. doi: 10.1113/jphysiol.1961.sp006806

Eyzaguirre, C., and Lewin, J. (1961). Chemoreceptor activity of the carotid body of the
cat. *J. Physiol.* 159, 222–237. doi: 10.1113/jphysiol.1961.sp006804

Eyzaguirre, C., and Zapata, P. (1984). Perspectives in carotid body research. *J. Appl.
Physiol. Respir. Environ. Exerc. Physiol.* 57, 931–957. doi:
10.1152/jappl.1984.57.4.931

Fatemian, M., Nieuwenhuijs, D. J., Teppema, L. J., Meinesz, S., van der Mey, A. G.,
Dahan, A., and Robbins, P. A. (2003). The respiratory response to carbon dioxide in
humans with unilateral and bilateral resections of the carotid bodies. *J. Physiol.* 549,
965-973. doi: 10.1113/jphysiol.2003.042259

Felder, R. B., Heesch, C. M., and Thames, M. D. (1983). Reflex modulation of carotid
sinus baroreceptor activity in the dog. *Am. J. Physiol.* 244, H437–H443. doi:
10.1152/ajpheart.1983.244.3.H437

Feldman-Goriachnik, R., and Hanani, M. (2019). The effects of sympathetic nerve
damage on satellite glial cells in the mouse superior cervical ganglion. *Auton.
Neurosci.* 221, 102584. doi: 10.1016/j.autneu.2019.102584

Fencl, V., Miller, T. B., and Pappenheimer, J. R. (1966). Studies on the respiratory
response to disturbances of acid-base balance, with deductions concerning the
ionic composition of cerebral interstitial fluid. *Am. J. Physiol.* 210, 459-472. doi:

10.1152/ajplegacy.1966.210.3.459

Flett, D. L., and Bell, C. (1991). Topography of functional subpopulations of neurons in the superior cervical ganglion of the rat. *J. Anat.* 177, 55–66.

Floyd, W. F., and Neil, E. (1952). The influence of the sympathetic innervation of the carotid bifurcation on chemoceptor and baroceptor activity in the cat. *Arch. Int. Pharmacodyn. Ther.* 91, 230–239.

Folgering, H., Ponte, J., and Sadig, T. (1982). Adrenergic mechanisms and chemoreception in the carotid body of the cat and rabbit. *J. Physiol.* 325, 1–21. doi: 10.1113/jphysiol.1982.sp014131

Forehand, C. J. (1985). Density of somatic innervation on mammalian autonomic ganglion cells is inversely related to dendritic complexity and preganglionic convergence. *J. Neurosci.* 5, 3403–3408. doi: 10.1523/JNEUROSCI.05-12-03403.1985

Forster, H. V. (2003). Plasticity in the control of breathing following sensory denervation. J. Appl. Physiol. (1985). 94, 784–794. doi: 10.1152/japplphysiol.00602.2002

Forster, H. V., and Smith, C. A. (2010). Contributions of central and peripheral chemoreceptors to the ventilatory response to CO_2/H^+. *J. Appl. Physiol. (1985).* 108, 989-994. doi: 10.1152/japplphysiol.01059.2009

Fung, M. L., Wang, W., Darnall, R. A., and St John, W. M. (1996). Characterization of ventilatory responses to hypoxia in neonatal rats. *Respir. Physiol.* 103, 57-66. doi: 10.1016/0034-5687(95)00077-1

Gallardo, E., Chiocchio, S. R., and Tramezzani, J. H. (1984). Sympathetic innervation of the median eminence. *Brain Res.* 290, 333–335. doi: 10.1016/0006-8993(84)90951-

x

Gao, L., Bonilla-Henao, V., García-Flores, P., Arias-Mayenco, I., Ortega-Sáenz, P., and López-Barneo, J. (2017). Gene expression analyses reveal metabolic specifications in acute O_2-sensing chemoreceptor cells. *J. Physiol.* 595, 6091–6120. doi: 10.1113/JP274684

Gao, L., Ortega-Sáenz, P., and López-Barneo, J. (2019). Acute oxygen sensing-Role of metabolic specifications in peripheral chemoreceptor cells. *Respir. Physiol. Neurobiol.* 265, 100–111. doi: 10.1016/j.resp.2018.08.007

García, J. B., Romeo, H. E., Basabe, J. C., and Cardinali, D. P. (1988). Effect of superior cervical ganglionectomy on insulin release by murine pancreas slices. *J. Auton. Nerv. Syst.* 22, 159–165. doi: 10.1016/0165-1838(88)90089-6

Gaston, B., May, W. J., Sullivan, S., Yemen, S., Marozkina, N. V., Palmer, L. A., Bates, J. N., and Lewis, S. J. (2014). Essential role of hemoglobin beta-93-cysteine in posthypoxia facilitation of breathing in conscious mice. *J. Appl. Physiol. (1985).* 116, 1290-1299. doi: 10.1152/japplphysiol.01050.2013

Gerard, M. W., and Billingsley, P. R. (1923). The innervation of the carotid body. Anat Rec. 25, 391–400.

Getsy, P. M., Davis, J., Coffee, G. A., May, W. J., Palmer, L. A., Strohl, K. P., and Lewis, S. J. (2014). Enhanced non-eupneic breathing following hypoxic, hypercapnic or hypoxic-hypercapnic gas challenges in conscious mice. *Respir. Physiol. Neurobiol.* 204, 147-159. doi: 10.1016/j.resp.2014.09.006

Getsy, P. M., Coffee, G. A., and Lewis, S. J. (2020). The Role of Carotid Sinus Nerve Input in the Hypoxic-Hypercapnic Ventilatory Response in Juvenile Rats. *Front. Physiol.* 11, 613786. doi: 10.3389/fphys.2020.613786

Getsy, P. M., Coffee, G. A., Hsieh, Y. H., and Lewis, S. J. (2021). The superior cervical ganglia modulate ventilatory responses to hypoxia independently of pre-ganglionic drive from the cervical sympathetic chain. *J. Appl. Physiol.* In press.

Getsy, P. M., Coffee, G. A., Hsieh, Y. H., Lewis, S. J. (2021). Loss of cervical sympathetic chain input to the superior cervical ganglia affects the ventilatory responses to hypoxic challenge in freely-moving C57BL6 mice. *Front. Physiol.* 12, 619688. doi: 10.3389/fphys.2021.619688

Gibbins, I. L. (1991). Vasomotor, pilomotor and secretomotor neurons distinguished by size and neuropeptide content in superior cervical ganglia of mice. *J. Auton. Nerv. Syst.* 34, 171-183. doi: 10.1016/0165-1838(91)90083-f

Gonsalves, S. F., Smith, E. J., Nolan, W .F., and Dutton, R. E. (1984). beta-Adrenoceptor blockade spares chemoreceptor responsiveness to hypoxia. *Brain Res.* 324, 349-353. doi: 10.1016/0006-8993(84)90047-7

González-Guerrero, P. R., Rigual, R., and González, C. (1993). Opioid peptides in the rabbit carotid body: identification and evidence for co-utilization and interactions with dopamine. *J. Neurochem.* 60, 1762-1768. doi: 10.1111/j.1471-4159.1993.tb13401.x

Gonzalez, C., Almaraz, L., Obeso, A., and Rigual, R. (1994). Carotid body chemoreceptors: from natural stimuli to sensory discharges. *Physiol. Rev.* 74, 829-898. doi: 10.1152/physrev.1994.74.4.829

Gourine, A. V., and Funk, G. D. (2017). On the existence of a central respiratory oxygen sensor. *J Appl Physiol (1985)*. 123, 1344–1349. doi: 10.1152/japplphysiol.00194.2017

Groeben, H., Meier, S., Tankersley, C. G., Mitzner, W., and Brown, R. H. (2005). Heritable and pharmacological influences on pauses and apneas in inbred mice during anesthesia and emergence. *Exp. Lung Res.* 31, 839–853. doi: 10.1080/01902140600586458

Gronda, E., Seravalle, G., Brambilla, G., Costantino, G., Casini, A., Alsheraei, A., Lovett, E. G., Mancia, G., and Grassi, G. (2014). Chronic baroreflex activation effects on sympathetic nerve traffic, baroreflex function, and cardiac haemodynamics in heart failure: a proof-of-concept study. *Eur J Heart Fail*. 16, 977–983. doi: 10.1002/ejhf.138

Guyenet, P. G., and Bayliss, D. A. (2015). Neural Control of Breathing and CO_2 Homeostasis. *Neuron*. 87, 946–961. doi: 10.1016/j.neuron.2015.08.001

Guyenet, P. G., Stornetta, R. L., Souza, G., Abbott, S., Shi, Y., and Bayliss, D. A. (2019). The Retrotrapezoid Nucleus: Central Chemoreceptor and Regulator of Breathing Automaticity. *Trends Neurosci*. 42, 807–824. doi: 10.1016/j.tins.2019.09.002

Hachmann, J. T., Grahn, P. J., Calvert, J. S., Drubach, D. I., Lee, K. H., and Lavrov, I. A. (2017). Electrical Neuromodulation of the Respiratory System After Spinal Cord Injury. *Mayo Clin. Proc.* 92, 1401-1414. doi: 10.1016/j.mayocp.2017.04.011

Han, F., and Strohl, K. P. (2000). Inheritance of ventilatory behavior in rodent models. *Respir. Physiol*. 121, 247-256. doi: 10.1016/s0034-5687(00)00132-8

Han, F., Subramanian, S., Dick, T. E., Dreshaj, I. A., and Strohl, K. P. (2001). Ventilatory behavior after hypoxia in C57BL/6J and A/J mice. *J. Appl. Physiol. (1985)*. 91, 1962-1970. doi: 10.1152/jappl.2001.91.5.1962

Han, F., Subramanian, S., Price, E. R., Nadeau, J., and Strohl, K. P. (2002). Periodic breathing in the mouse. *J. Appl. Physiol. (1985)*. 92, 1133-1140. doi: 10.1152/japplphysiol.00785.2001

Hanson, G., Jones, L., and Fidone, S. (1986). Effects of hypoxia on neuropeptide levels in the rabbit superior cervical ganglion. *J. Neurobiol*. 17, 51-54. doi: 10.1002/neu.480170106

He, L., Chen, J., Dinger, B., and Fidone, S. (2000). Characteristics of carotid body chemosensitivity in the mouse. Baseline studies for future experiments with knockout animals. *Adv. Exp. Med. Biol*. 475, 697-704. doi: 10.1007/0-306-46825-5_69

He, L., Chen, J., Dinger, B., Sanders, K., Sundar, K., Hoidal, J., and Fidone, S. (2002). Characteristics of carotid body chemosensitivity in NADPH oxidase-deficient mice. *Am. J. Physiol. Cell Physiol*. 282, C27–C33. doi: 10.1152/ajpcell.2002.282.1.C27

He, L., Chen, J., Dinger, B., Sanders, K., Sundar, K., Hoidal, J., and Fidone, S. (2003). Carotid body chemoreceptor activity in mice deficient in selected subunits of NADPH oxidase. *Adv. Exp. Med. Biol*. 536, 41-46. doi: 10.1007/978-1-4419-9280-2_5

Heinert, G., Paterson, D. J., Bisgard, G. E., Xia, N., Painter, R., and Nye, P. C. (1995). The excitation of carotid body chemoreceptors of the cat by potassium and noradrenaline. *Adv. Exp. Med. Biol*. 393, 323–330. doi: 10.1007/978-1-4615-1933-

1_61

Hellström S. (1975). Morphometric studies of dense-cored vesicles in type I cells of rat carotid body. *J Neurocytol.* 4, 77–86. doi: 10.1007/BF01099097

Henderson, F., May, W. J., Gruber, R. B., Young, A. P., Palmer, L. A., Gaston, B., and Lewis, S. J. (2013). Low-dose morphine elicits ventilatory excitant and depressant responses in conscious rats: Role of peripheral μ-opioid receptors. *Open J. Mol. Integr. Physiol.* 3, 111–124. doi: 10.4236/ojmip.2013.33017

Henderson, F., May, W. J., Gruber, R. B., Discala, J. F., Puskovic, V., Young, A. P., Baby, S. M., and Lewis, S. J. (2014). Role of central and peripheral opiate receptors in the effects of fentanyl on analgesia, ventilation and arterial blood-gas chemistry in conscious rats. *Respir. Physiol. Neurobiol.* 191, 95–105. doi: 10.1016/j.resp.2013.11.005

Heutink, M., Post, M. W., Luthart, P., Schuitemaker, M., Slangen, S., Sweers, J., Vlemmix, L., and Lindeman, E. (2014). Long-term outcomes of a multidisciplinary cognitive behavioural programme for coping with chronic neuropathic spinal cord injury pain. *J. Rehabil. Med.* 46, 540-545. doi: 10.2340/16501977-1798

Hilaire, G., Viemari, J. C., Coulon, P., Simonneau, M., and Bévengut, M. (2004). Modulation of the respiratory rhythm generator by the pontine noradrenergic A5 and A6 groups in rodents. *Respir Physiol Neurobiol.* 143, 187–197. doi: 10.1016/j.resp.2004.04.016

Hisa, Y., Koike, S., Tadaki, N., Bamba, H., Shogaki, K., and Uno, T. (1999). Neurotransmitters and neuromodulators involved in laryngeal innervation. *Ann.*

Otol. Rhinol. Laryngol. Suppl. 178, 3-14. doi: 10.1177/00034894991080s702

Hitzig, B. M., Perng, W. C., Burt, T., Okunieff, P., and Johnson, D. C. (1994). 1H-NMR measurement of fractional dissociation of imidazole in intact animals. *Am J Physiol*. 266, R1008–R1015. doi: 10.1152/ajpregu.1994.266.3.R1008

Hodson, E. J., Nicholls, L. G., Turner, P. J., Llyr, R., Fielding, J. W., Douglas, G., Ratnayaka, I., Robbins, P. A., Pugh, C. W., Buckler, K. J., Ratcliffe, P. J., and Bishop, T. (2016). Regulation of ventilatory sensitivity and carotid body proliferation in hypoxia by the PHD2/HIF-2 pathway. *J Physiol*. 594, 1179–1195. doi: 10.1113/JP271050

Hughes-Davis, E. J., Cogen, J. P., Jakowec, M. W., Cheng, H. W., Grenningloh, G., Meshul, C. K., and McNeill, T. H. (2005). Differential regulation of the growth-associated proteins GAP-43 and superior cervical ganglion 10 in response to lesions of the cortex and substantia nigra in the adult rat. *Neuroscience*. 135, 1231–1239. doi: 10.1016/j.neuroscience.2005.07.017

Ichikawa, H., and Helke, C. J. (1993). Distribution, origin and plasticity of galanin-immunoreactivity in the rat carotid body. *Neuroscience*. 52, 757–767. doi: 10.1016/0306-4522(93)90424-e

Ichikawa, H. (2002). Innervation of the carotid body: Immunohistochemical, denervation, and retrograde tracing studies. *Microsc. Res. Tech*. 59, 188–195. doi: 10.1002/jemt.10193

Inoue, K. (1975). *Kaibogaku Zasshi*. 50, 1–14.

Iturriaga, R., Mokashi, A., and Lahiri, S. (1993). Dynamics of carotid body responses in vitro in the presence of CO2-HCO3-: role of carbonic anhydrase. *J Appl Physiol (1985)*. 75, 1587–1594. doi: 10.1152/jappl.1993.75.4.1587

Iturriaga, R. and Alcayaga, J. (2004). Neurotransmission in the carotid body: transmitters and modulators between glomus cells and petrosal ganglion nerve terminals. *Brain Res Rev*. 47, 46–53. doi: 10.1016/j.brainresrev.2004.05.007

Iturriaga, R., Andrade, D. C., and Del Rio, R. (2014). Enhanced carotid body chemosensory activity and the cardiovascular alterations induced by intermittent hypoxia. *Front Physiol*. 5, 468. doi: 10.3389/fphys.2014.00468

Ivanusic, J. J., Wood, R. J., and Brock, J. A. (2013). Sensory and sympathetic innervation of the mouse and guinea pig corneal epithelium. *J. Comp. Neurol*. 521, 877–893. doi: 10.1002/cne.23207

Jiang, B., Huang, Z. J., Morales, B., and Kirkwood, A. (2005). Maturation of GABAergic transmission and the timing of plasticity in visual cortex. *Brain Res. Rev*. 50, 126-133. doi: 10.1016/j.brainresrev.2005.05.007

Jobling, P., and Gibbins, I. L. (1999). Electrophysiological and morphological diversity of mouse sympathetic neurons. *J. Neurophysiol*. 82, 2747–2764. doi: 10.1152/jn.1999.82.5.2747

Joseph, V., and Pequignot, J. M. (2009). Breathing at high altitude. *Cell Mol Life Sci*. 66, 3565–3573. doi: 10.1007/s00018-009-0143-y

Kåhlin, J., Eriksson, L. I., Ebberyd, A., and Fagerlund, M. J. (2010). Presence of nicotinic, purinergic and dopaminergic receptors and the TASK-1 K$^+$-channel in the mouse

carotid body. *Respir. Physiol. Neurobiol.* 172, 122–128. doi:
10.1016/j.resp.2010.05.001

Kamath, G. S., Rorie, D. K., and Tyce, G. M. (1993). Altered release and metabolism of
norepinephrine in superfused canine saphenous veins in the presence of halothane
and hypoxia. *Anesthesiology.* 78, 553-561. doi: 10.1097/00000542-199303000-
00019.

Karlsen, A. S., Rath, M. F., Rohde, K., Toft, T., and Møller, M. (2013). Developmental and
diurnal expression of the synaptosomal-associated protein 25 (Snap25) in the rat
pineal gland. *Neurochem. Res.* 38, 1219–1228. doi: 10.1007/s11064-012-0918-7

Kasa, P., Dobo, E., and Wolff, J. R. (1991). Are there cholinergic through-fibers in the
superior cervical ganglion of the mouse?. *Histochemistry.* 96, 261–263. doi:
10.1007/BF00271545

Kawaja, M. D., and Crutcher, K. A. (1997). Sympathetic axons invade the brains of mice
overexpressing nerve growth factor. *J. Comp. Neurol.* 383, 60–72.

Kidd, G. J., and Heath, J. W. (1988). Double myelination of axons in the sympathetic
nervous system of the mouse. I. Ultrastructural features and distribution. *J.
Neurocytol.* 17, 245–261. doi: 10.1007/BF01674211

Kikuta, S., Iwanaga, J., Kusukawa, J., and Tubbs, R. S. (2019). Carotid Sinus Nerve: A
Comprehensive Review of Its Anatomy, Variations, Pathology, and Clinical
Applications. *World Neurosurg.* 127, 370-374. doi: 10.1016/j.wneu.2019.04.064

Kirby, G. C., and McQueen, D. S. (1986). Characterization of opioid receptors in the cat carotid body involved in chemosensory depression in vivo. *Br J Pharmacol.* 88, 889–898. doi: 10.1111/j.1476-5381.1986.tb16263.x

Kiwull, P., Kiwull-Schöne, H., and Klatt, W. (1976). Interaction of central and peripheral respiratory drives: differentiation between the role of stimuli and afferents. In *Acid-Base Homeostasis of Brain Extracellular Fluid and the Respiratory Control System.* ed. Loeschcke, H. H. 146–156. Stuttgart: Thieme

Kiwull-Schöne, H., Kiwull, P., Mückenhoff, K., and Both, W. (1976). The role of carotid chemoreceptors in the regulation of arterial oxygen transport under hypoxia with and without hypercapnia. *Adv. Exp. Med. Biol.* 75, 469-476. doi: 10.1007/978-1-4684-3273-2_55

Kline, D. D., and Prabhakar, N. R. (2000). Peripheral chemosensitivity in mutant mice deficient in nitric oxide synthase. *Adv. Exp. Med. Biol.* 475, 571–579. doi: 10.1007/0-306-46825-5_55

Kline, D. D., Buniel, M. C., Glazebrook, P., Peng, Y. J., Ramirez-Navarro, A., Prabhakar, N. R., and Kunze, D. L. (2005). Kv1.1 deletion augments the afferent hypoxic chemosensory pathway and respiration. *J. Neurosci.* 25, 3389–3399. doi: 10.1523/JNEUROSCI.4556-04.2005

Kogo, N., and Arita, H. (1990). In vivo study on medullary H(+)-sensitive neurons. *J Appl Physiol (1985).* 69, 1408–1412. doi: 10.1152/jappl.1990.69.4.1408

Kostreva, D. R., Palotas, G. L., and Kampine, J. P. (1984). Vagal inhibition of respiratory-

linked efferent activity in the carotid sinus nerve. *Am J Physiol*. 247, R681–R686.

doi: 10.1152/ajpregu.1984.247.4.R681

Kou, Y. R., Ernsberger, P., Cragg, P. A., Cherniack, N. S., and Prabhakar, N. R. (1991). Role

of alpha 2-adrenergic receptors in the carotid body response to isocapnic hypoxia.

Respir. Physiol. 83, 353–364. doi: 10.1016/0034-5687(91)90054-m

Krieger, D. T., Hauser, H., Liotta, A., and Zelenetz, A. (1976). Circadian periodicity of

epidermal growth factor and its abolition by superior cervical ganglionectomy.

Endocrinology. 99, 1589–1596. doi: 10.1210/endo-99-6-1589

Kumar, P., and Prabhakar, N. R. (2012). Peripheral chemoreceptors: function and

plasticity of the carotid body. *Compr Physiol*. 2, 141–219. doi:

10.1002/cphy.c100069

Kummer, W., Fischer, A., Kurkowski, R., and Heym, C. (1992). The sensory and

sympathetic innervation of guinea-pig lung and trachea as studied by retrograde

neuronal tracing and double-labelling immunohistochemistry. *Neuroscience*. 49,

715–737. doi: 10.1016/0306-4522(92)90239-x

Kurz, T., Richardt, G., Hagl, S., Seyfarth, M., and Schömig, A. (1995). Two different

mechanisms of noradrenaline release during normoxia and simulated ischemia in

human cardiac tissue. *J. Mol. Cell Cardiol*. 27, 1161–1172. doi: 10.1016/0022-

2828(95)90052-7

Kurz, T., Richardt, G., Seyfarth, M., and Schömig, A. (1996). Nonexocytotic noradrenaline

release induced by pharmacological agents or anoxia in human cardiac tissue. *Naunyn Schmiedebergs Arch. Pharmacol.* 354, 7–16. doi: 10.1007/BF00168700

Kusakabe, T., Hayashida, Y., Matsuda, H., Gono, Y., Powell, F. L., Ellisman, M. H., Kawakami, T., and Takenaka, T. (1998). Hypoxic adaptation of the peptidergic innervation in the rat carotid body. *Brain Res.* 806, 165–174. doi: 10.1016/s0006-8993(98)00742-2

Kusakabe, T., Hirakawa, H., Matsuda, H., Kawakami, T., Takenaka, T., and Hayashida, Y. (2003). Peptidergic innervation in the rat carotid body after 2, 4, and 8 weeks of hypocapnic hypoxic exposure. *Histol Histopathol.* 18, 409–418. doi: 10.14670/HH-18.409

Kusakabe, T., Hirakawa, H., Oikawa, S., Matsuda, H., Kawakami, T., Takenaka, T., and Hayashida, Y. (2004). Morphological changes in the rat carotid body 1, 2, 4, and 8 weeks after the termination of chronically hypocapnic hypoxia. *Histol Histopathol.* 19, 1133–1140. doi: 10.14670/HH-19.1133

Laferrière, A., Colin-Durand, J., and Moss, I. R. (2005). Ontogeny of respiratory sensitivity and tolerance to the mu-opioid agonist fentanyl in rat. *Brain res. Develop. brain res.*, 156, 210–217. doi: 10.1016/j.devbrainres.2005.03.002

Lahiri, S., Pokorski, M., and Davies, R. O. (1981). Augmentation of carotid body chemoreceptor responses by isoproterenol in the cat. *Respir. Physiol.* 44, 351–364. doi: 10.1016/0034-5687(81)90029-3

Lahiri, S., Matsumoto, S., and Mokashi, A. (1986). Responses of ganglioglomerular nerve activity to respiratory stimuli in the cat. *J. Appl. Physiol. (1985).* 60, 391–397. doi:

10.1152/jappl.1986.60.2.391

Lai, C. H., Ma, C. W., Lai, S. K., Han, L., Wong, H. M., Yeung, K. W., Shum, D. K., and Chan, Y. S. (2016). Maturation of glutamatergic transmission in the vestibulo-olivary pathway impacts on the registration of head rotational signals in the brainstem of rats. *Brain Struct. Funct.* 221, 217-38. doi: 10.1007/s00429-014-0903-9

Lamb, T. W. (1966). Ventilatory responses to intravenous and inspired carbon dioxide in anesthetized cats. *Respir Physiol.* 2, 99–104. doi: 10.1016/0034-5687(66)90041-7

Lewis, J. C., and Burton, P. R. (1977). Ultrastructural studies of the superior cervical trunk of the mouse: distribution, cytochemistry and stability of fibrous elements in preganglionic fibers. *J. Comp. Neurol.* 171, 605–618. doi: 10.1002/cne.901710411

Li, A., and Nattie, E. (2002). CO2 dialysis in one chemoreceptor site, the RTN: stimulus intensity and sensitivity in the awake rat *Respir Physiol Neurobiol.* 133, 11–22. doi: 10.1016/s1569-9048(02)00134-9

Li, A., and Nattie, E. (2008). Serotonin transporter knockout mice have a reduced ventilatory response to hypercapnia (predominantly in males) but not to hypoxia. *J Physiol.* 586, 2321–2329. doi: 10.1113/jphysiol.2008.152231

Li, A., Zhou, S., and Nattie, E. (2006). Simultaneous inhibition of caudal medullary raphe and retrotrapezoid nucleus decreases breathing and the CO2 response in conscious rats. *J Physiol.* 577, 307–318. doi: 10.1113/jphysiol.2006.114504

Libet, B. (1970). Generation of slow inhibitory and excitatory postsynaptic potentials. *Fedn. Proc.* 29, 1945-1956.

Libet, B., and Tosaka, T. (1970). Dopamine as a synaptic transmitter and modulator in

sympathetic ganglia: a different mode of synaptic action. Proc. Natl. Acad. Sci. USA.

67, 667-673. doi: 10.1073/pnas.67.2.667

Libet, B., and Owman, C. (1974). Concomitant changes in formaldehyde-induced

fluorescence of dopamine interneurones and in slow inhibitory post-synaptic

potentials of the rabbit superior cervical ganglion, induced by stimulation of the

preganglionic nerve or by a muscarinic agent. *J. Physiol*. 237, 635-662. doi:

10.1113/jphysiol.1974.sp010502

Limberg, J. K. (2018). Glucose, insulin, and the carotid body chemoreceptors in

humans. *Physiol Genomics*. 50, 504–509. doi: 10.1152/physiolgenomics.00032.2018

Lindborg, J. A., Niemi, J. P., Howarth, M. A., Liu, K. W., Moore, C. Z., Mahajan, D., and

Zigmond, R. E. (2018). Molecular and cellular identification of the immune response

in peripheral ganglia following nerve injury. *J. Neuroinflammation*. 15, 192. doi:

10.1186/s12974-018-1222-5

Linton, R. A., Miller, R., and Cameron, I. R. (1976). Ventilatory response to CO2

inhalation and intravenous infusion of hypercapnic blood. *Respir Physiol*. 26, 383–

394. doi: 10.1016/0034-5687(76)90008-6

Little, G. J., and Heath, J. W. (1994). Morphometric analysis of axons myelinated during

adult life in the mouse superior cervical ganglion. *J. Anat*. 184, 387–398.

Liu, P. W., and Bean, B. P. (2014). Kv2 channel regulation of action potential

repolarization and firing patterns in superior cervical ganglion neurons and

hippocampal CA1 pyramidal neurons. *J. Neurosci*. 34, 4991–5002. doi:

10.1523/JNEUROSCI.1925-13.2014

Liu, Q., Lowry, T. F., and Wong-Riley, M. T. (2006). Postnatal changes in ventilation during normoxia and acute hypoxia in the rat: implication for a sensitive period. *J. Physiol.* 577, 957–970. doi: 10.1113/jphysiol.2006.121970

Llados, F., and Zapata, P. (1978). Effects of adrenoceptor stimulating and blocking agents on carotid body chemosensory inhibition. *J. Physiol.* 274, 501–509. doi: 10.1113/jphysiol.1978.sp012163

Llewellyn-Smith, I. J., Arnolda, L. F., Pilowsky, P. M., Chalmers, J. P., and Minson, J. B. (1998). GABA- and glutamate-immunoreactive synapses on sympathetic preganglionic neurons projecting to the superior cervical ganglion. *J. Auton. Nerv. Syst.* 71, 96–110. doi: 10.1016/s0165-1838(98)00069-1

López-Barneo, J., Ortega-Sáenz, P., Pardal, R., Pascual, A., and Piruat, J. I. (2008). Carotid body oxygen sensing. *Eur. Respir. J.* 32, 1386–1398. doi: 10.1183/09031936.00056408

López-Barneo, J., Ortega-Sáenz, P., González-Rodríguez, P., Fernández-Agüera, M. C., Macías, D., Pardal, R., and Gao, L. (2016). Oxygen-sensing by arterial chemoreceptors: Mechanisms and medical translation. *Mol Aspects Med.* 47-48, 90–108. doi: 10.1016/j.mam.2015.12.002

López-Barneo, J., González-Rodríguez, P., Gao, L., Fernández-Agüera, M. C., Pardal, R., and Ortega-Sáenz, P. (2016). Oxygen sensing by the carotid body: mechanisms and role in adaptation to hypoxia. *Am J Physiol Cell Physiol. Cell physiology*, 310, C629–C642. doi: 10.1152/ajpcell.00265.2015

López-López, J., González, C., Ureña, J., and López-Barneo, J. (1989). Low pO2 selectively inhibits K channel activity in chemoreceptor cells of the mammalian carotid body. *J Gen Physiol*. 93, 1001–1015. doi: 10.1085/jgp.93.5.1001

Lowry, T. F., Forster, H. V., Pan, L. G., Korducki, M. A., Probst, J., Franciosi, R. A., and Forster, M. (1999a). Effect of carotid body denervation on breathing in neonatal goats. *J. Appl. Physiol.* (1985). 87, 1026–1034. doi: 10.1152/jappl.1999.87.3.1026

Lowry, T. F., Forster, H. V., Pan, L. G., Serra, A., Wenninger, J., Nash, R., Sheridan, D., and Franciosi, R. A. (1999b). Effects on breathing of carotid body denervation in neonatal piglets. *J. Appl. Physiol.* (1985). 87, 2128–2135. doi: 10.1152/jappl.1999.87.6.2128

Lujan, H. L., and DiCarlo, S. E. (2007). T5 spinal cord transection increases susceptibility to reperfusion-induced ventricular tachycardia by enhancing sympathetic activity in conscious rats. *Am. J. Physiol. Heart Circ. Physiol.* 293, H3333–H3339. doi: 10.1152/ajpheart.01019.2007

Lujan, H. L., Chen, Y., and Dicarlo, S. E. (2009). Paraplegia increased cardiac NGF content, sympathetic tonus, and the susceptibility to ischemia-induced ventricular tachycardia in conscious rats. *Am. J. Physiol. Heart Circ. Physiol.* 296, H1364–H1372. doi: 10.1152/ajpheart.01286.2008

Lujan, H. L., Palani, G., and DiCarlo, S. E. (2010). Structural neuroplasticity following T5 spinal cord transection: increased cardiac sympathetic innervation density and SPN arborization. *Am. J. Physiol. Regul. Integr. Comp. Physiol.* 299, R985–R995. doi: 10.1152/ajpregu.00329.2010

Lujan, H. L., Janbaih, H., and DiCarlo, S. E. (2012). Dynamic interaction between the

heart and its sympathetic innervation following T5 spinal cord transection. *J. Appl.

Physiol. (1985)*. 113, 1332–1341. doi: 10.1152/japplphysiol.00522.2012

Lujan, H. L., Janbaih, H., and DiCarlo, S. E. (2014). Structural remodeling of the heart and

its premotor cardioinhibitory vagal neurons following T(5) spinal cord transection. *J.

Appl. Physiol. (1985)*. 116, 1148–1155. doi: 10.1152/japplphysiol.01285.2013

Macías, D., Fernández-Agüera, M. C., Bonilla-Henao, V., and López-Barneo, J. (2014).

Deletion of the von Hippel-Lindau gene causes sympathoadrenal cell death and

impairs chemoreceptor-mediated adaptation to hypoxia. *EMBO Mol Med*. 6, 1577–

1592. doi: 10.15252/emmm.201404153

Macias, D., Cowburn, A. S., Torres-Torrelo, H., Ortega-Sáenz, P., López-Barneo, J., and

Johnson, R. S. (2018). HIF-2α is essential for carotid body development and

function. *eLife*. 7, e34681. doi: 10.7554/eLife.34681

Majcherczyk, S., Chrucielewski, L., and Trzebski, A. (1974). Effect of stimulation of

carotid body chemoreceptors upon ganglioglomerular nerve activity and on

chemoreceptor discharges in contralateral sinus nerve. *Brain Res*. 76, 167–170. doi:

10.1016/0006-8993(74)90524-1

Majcherczyk, S., Coleridge, J. C., Coleridge, H. M., Kaufman, M. P., and Baker, D. G.

(1980). Carotid sinus nerve efferents: properties and physiological significance. *Fed.

Proc*. 39, 2662-2667.

Maklad, A., Quinn, T., and Fritzsch, B. (2001). Intracranial distribution of the sympathetic

system in mice: DiI tracing and immunocytochemical labeling. *Anat. Rec*. 263, 99-

111. doi: 10.1002/ar.1083

Marcus, N. J., Li, Y. L., Bird, C. E., Schultz, H. D., and Morgan, B. J. (2010). Chronic intermittent hypoxia augments chemoreflex control of sympathetic activity: role of the angiotensin II type 1 receptor. *Respir Physiol Neurobiol*. 171, 36–45. doi: 10.1016/j.resp.2010.02.003

Marjanovic, M., Elliott, A. C., and Dawson, M. J. (1998). The temperature dependence of intracellular pH in isolated frog skeletal muscle: lessons concerning the Na(+)-H+ exchanger. *J Membr Biol*. 161, 215–225. doi: 10.1007/s002329900328

Martin-Body, R. L., Robson, G. J., and Sinclair, J. D. (1985). Respiratory effects of sectioning the carotid sinus glossopharyngeal and abdominal vagal nerves in the awake rat. *J. Physiol*. 361, 35–45. doi: 10.1113/jphysiol.1985.sp015631

Martin-Body, R. L., Robson, G. J., and Sinclair, J. D. (1986). Restoration of hypoxic respiratory responses in the awake rat after carotid body denervation by sinus nerve section. *J. Physiol*. 380, 61–73. doi: 10.1113/jphysiol.1986.sp016272

Martinez-Pinna, J., Soriano, S., Tudurí, E., Nadal, A., and de Castro, F. (2018). A Calcium-Dependent Chloride Current Increases Repetitive Firing in Mouse Sympathetic Neurons. *Front. Physiol*. 9, 508. doi: 10.3389/fphys.2018.00508

Massari, V. J., Shirahata, M., Johnson, T. A., and Gatti, P. J. (1996). Carotid sinus nerve terminals which are tyrosine hydroxylase immunoreactive are found in the commissural nucleus of the tractus solitarius. *J Neurocytol*. 25, 197–208. doi: 10.1007/BF02284796

Mathew, T. C. (2007). Scanning electron microscopic observations on the third ventricular floor of the rat following cervical sympathectomy. *Folia Morphol.*

(Warsz). 66, 94–99.

Matsumoto, S., Nagao, T., Ibi, A., and Nakajima, T. (1980). Effects of carotid body

chemoreceptor stimulation by dopamine on ventilation. *Arch. Int. Pharmacodyn.*

Ther. 245, 145–155.

Matsumoto, S., Ibi, A., Nagao, T., and Nakajima, T. (1981). Effects of carotid body

chemoreceptor stimulation by norepinephrine, epinephrine and tyramine on

ventilation in the rabbit. *Arch. Int. Pharmacodyn. Ther.* 252, 152-161.

Matsumoto, S., Mokashi, A., and Lahiri, S. (1986). Influence of ganglioglomerular nerve

on carotid chemoreceptor activity in the cat. *J. Auton. Nerv. Syst.* 15, 7-20. doi:

10.1016/0165-1838(86)90075-5

Matsumoto, S., Mokashi, A., and Lahiri, S. (1987). Ganglioglomerular nerves respond to

moderate hypoxia independent of peripheral chemoreceptors in the cat. *J. Auton.*

Nerv. Syst. 19, 219–228. doi: 10.1016/0165-1838(87)90068-3

May, W. J., Gruber, R. B., Discala, J. F., Puskovic, V., Henderson, F., Palmer, L. A., and

Lewis, S. J. (2013a). Morphine has latent deleterious effects on the ventilatory

responses to a hypoxic challenge. *Open J. Mol. Integr. Physiol.* 3, 166–180. doi:

10.4236/ojmip.2013.34022

May, W. J., Henderson, F., Gruber, R. B., Discala, J. F., Young, A. P., Bates, J. N., Palmer, L.

A., and Lewis, S. J. (2013b). Morphine has latent deleterious effects on the

ventilatory responses to a hypoxic-hypercapnic challenge. *Open J. Mol. Integr.*

Physiol. 3, 134–145. doi: 10.4236/ojmip.2013.33019

Maxová, H., and Vízek, M. (2001). Biphasic ventilatory response to hypoxia in

unanesthetized rats. Physiol. Res. 50, 91–96.

McCooke, H. B. and Hanson, M. A. (1985). Respiration of conscious kittens in acute

hypoxia and effect of almitrine bismesylate. *J. Appl. Physiol.* (1985). 59, 18–23. doi:

10.1152/jappl.1985.59.1.18

McDonald, D. M., and Mitchell, R. A. (1975). The innervation of glomus cells, ganglion

cells and blood vessels in the rat carotid body: a quantitative ultrastructural

analysis. *J Neurocytol.* 4, 177–230.

McDonald, D.M., and Blewett, R.W. (1981). Location and size of carotid body-like organs

(paraganglia) revealed in rats by the permeability of blood vessels to Evans blue

dye. *J. Neurocytol.* 10, 607-643. doi: 10.1007/BF01262593

McDonald, D. M., and Mitchell, R. A. (1981). The neural pathway involved in "efferent

inhibition" of chemoreceptors in the cat carotid body. *J. Comp. Neurol.* 201, 457-

476. doi: 10.1002/cne.902010310

McDonald, D. M. (1983a). Morphology of the rat carotid sinus nerve. I. Course,

connections, dimensions and ultrastructure. *J. Neurocytol.* 12, 345–372. doi:

10.1007/BF01159380

McDonald, D.M. (1983b). Morphology of the rat carotid sinus nerve. II. Number and size

of axons. *J. Neurocytol.* 12, 373-392. doi: 10.1007/BF01159381

McLachlan, E. M. (2007). Diversity of sympathetic vasoconstrictor pathways and their

plasticity after spinal cord injury. *Clin. Auton. Res.* 17, 6–12. doi:

org/10.1007/s10286-006-0394-8

McQueen, D. S., and Ribeiro, J. A. (1980). Inhibitory actions of methionine-enkephalin

and morphine on the cat carotid chemoreceptors. *Br J Pharmacol*. 71, 297–305. doi:

10.1111/j.1476-5381.1980.tb10939.x

McQueen, D. S., Evrard, Y., Gordon, B. H., and Campbell, D. B. (1989). Ganglioglomerular

nerves influence responsiveness of cat carotid body chemoreceptors to almitrine. *J.

Auton. Nerv. Syst.* 27, 57–66. doi: 10.1016/0165-1838(89)90129-x

Meckler, R. L., and Weaver, L. C. (1985). Splenic, renal, and cardiac nerves have unequal

dependence upon tonic supraspinal inputs. *Brain Res.* 338, 123–135. doi:

10.1016/0006-8993(85)90254-9

Messier, M. L., Li, A., and Nattie, E. E. (2004). Inhibition of medullary raphe serotonergic

neurons has age-dependent effects on the CO2 response in newborn piglets. *J Appl

Physiol (1985)*. 96, 1909–1919. doi: 10.1152/japplphysiol.00805.2003

Mills, E., Smith, P. G., Slotkin, T. A., and Breese, G. (1978). Role of carotid body

catecholamines in chemoreceptor function. *Neuroscience.* 3, 1137–1146. doi:

10.1016/0306-4522(78)90134-3

Milsom, W. K., and Sadig, T. (1983). Interaction between norepinephrine and hypoxia on

carotid body chemoreception in rabbits. *J. Appl. Physiol. Respir. Environ. Exerc.

Physiol.* 55, 1893–1898. doi: 10.1152/jappl.1983.55.6.1893

Mir, A. K., Al-Neamy, K., Pallot, D. J., and Nahorski, S. R. (1982). Catecholamines in the

carotid body of several mammalian species: effects of surgical and chemical

sympathectomy. *Brain Res.* 252, 335–342. doi: 10.1016/0006-8993(82)90401-2

Mitchell, J. H., and McCloskey, D. I. (1974). Chemoreceptor responses to sympathetic

stimulation and changes in blood pressure. *Respir Physiol*. 20, 297–302. doi:

10.1016/0034-5687(74)90026-7

Mitchell, G. S., and Johnson, S. M. (2003). Neuroplasticity in respiratory motor control. *J.

Appl. Physiol*. (1985). 94, 358–374. doi: 10.1152/japplphysiol.00523.2002

Mitsuoka, K., Miwa, Y., Kikutani, T., and Sato, I. (2018). Localization of CGRP and VEGF

mRNAs in the mouse superior cervical ganglion during pre- and postnatal

development. *Eur. J. Histochem*. 62, 2976. doi: 10.4081/ejh.2018.2976

Moore, M. W., Chai, S., Gillombardo, C. B., Carlo, A., Donovan, L. M., Netzer, N., and

Strohl, K. P. (2012). Two weeks of buspirone protects against posthypoxic

ventilatory pauses in the C57BL/6J mouse strain. *Respir. Physiol. Neurobiol*. 183, 35-

40. doi: 10.1016/j.resp.2012.05.001

Moore, M. W., Akladious, A., Hu, Y., Azzam, S., Feng, P., and Strohl, K. P. (2014). Effects

of orexin 2 receptor activation on apnea in the C57BL/6J mouse. *Respir. Physiol.

Neurobiol*. 200, 118-125. doi: 10.1016/j.resp.2014.03.014

Mouradian, G. C., Forster, H. V., and Hodges, M. R. (2012). Acute and chronic effects of

carotid body denervation on ventilation and chemoreflexes in three rat strains. *J.

Physiol*. 590, 3335–3347. doi: 10.1113/jphysiol.2012.234658

Narkiewicz, K., Ratcliffe, L. E., Hart, E. C., Briant, L. J., Chrostowska, M., Wolf, J., Szyndler,

A., Hering, D., Abdala, A. P., Manghat, N., Burchell, A. E., Durant, C., Lobo, M. D.,

Sobotka, P. A., Patel, N. K., Leiter, J. C., Engelman, Z. J., Nightingale, A. K., and Paton,

J. F. (2016). Unilateral Carotid Body Resection in Resistant Hypertension: A Safety and Feasibility Trial. *JACC Basic Transl Sci*. 1, 313–324. doi:

10.1016/j.jacbts.2016.06.004

Nattie, E. E., and Li, A. (1994). Retrotrapezoid nucleus lesions decrease phrenic activity

and CO2 sensitivity in rats. *Respir Physiol*. 97, 63–77. doi: 10.1016/0034-

5687(94)90012-4

Nattie, E. E., and Li, A. (1996). Central chemoreception in the region of the ventral

respiratory group in the rat. *J Appl Physiol (1985)*. 81, 1987–1995. doi:

10.1152/jappl.1996.81.5.1987

Nattie, E. (1998). Control and disturbances of cerebrospinal fluid pH. In: Kaila, K.; Silver,

RB., editors. pH and Brain Function. New York: Wiley-Liss, 629–650.

Nattie E. (2006). Why do we have both peripheral and central chemoreceptors?. *J. Appl.

Physiol. (1985)*. 100, 9–10. doi: 10.1152/japplphysiol.01097.2005

Nattie, E., and Li, A. (2012). Central chemoreceptors: locations and functions. *Compr

Physiol*. 2, 221–254. doi: 10.1002/cphy.c100083

Neistadt, A., and Schwartz, S. I. (1967). Effects of electrical stimulation of the carotid

sinus nerve in reversal of experimentally induced hypertension. *Surgery*. 61, 923–

931.

Niewinski, P., Janczak, D., Rucinski, A., Tubek, S., Engelman, Z. J., Piesiak, P., Jazwiec, P.,

Banasiak, W., Fudim, M., Sobotka, P. A., Javaheri, S., Hart, E. C., Paton, J. F., and

Ponikowski, P. (2017). Carotid body resection for sympathetic modulation in systolic

heart failure: results from first-in-man study. *Eur J Heart Fail*. 19, 391–400. doi:

10.1002/ejhf.641

Nunes, A. R., Batuca, J. R., and Monteiro, E. C. (2010). Acute hypoxia modifies cAMP

levels induced by inhibitors of phosphodiesterase-4 in rat carotid bodies, carotid

arteries and superior cervical ganglia. *Br. J. Pharmacol.* 159, 353-361. doi:

10.1111/j.1476-5381.2009.00534.x

Nunes, A. R., Sample, V., Xiang, Y. K., Monteiro, E. C., Gauda, E., and Zhang, J. (2012).

Effect of oxygen on phosphodiesterases (PDE) 3 and 4 isoforms and PKA activity in

the superior cervical ganglia. *Adv. Exp. Med. Biol.* 758, 287–294. doi: 10.1007/978-

94-007-4584-1_39

Nurse, C. A. (1990). Carbonic anhydrase and neuronal enzymes in cultured glomus cells

of the carotid body of the rat. *Cell Tissue Res.* 261, 65–71. doi: 10.1007/BF00329439

Nurse, C.A. (2014). Synaptic and paracrine mechanisms at carotid body arterial

chemoreceptors. *J. Physiol.* 592, 3419-3426. doi: 10.1113/jphysiol.2013.269829

Nurse, C. A., and Piskuric, N. A. (2013). Signal processing at mammalian carotid body

chemoreceptors. *Semin Cell Dev Biol.* 24, 22–30. doi: 10.1016/j.semcdb.2012.09.006

Nurse C. A. (2005). Neurotransmission and neuromodulation in the chemosensory

carotid body. *Auton. Neurosci.* 120, 1-9. doi: 10.1016/j.autneu.2005.04.008

Nurse C. A. (2010). Neurotransmitter and neuromodulatory mechanisms at peripheral

arterial chemoreceptors. *Exp. Physiol.* 95, 657–667. doi:

10.1113/expphysiol.2009.049312

Oh, E. J., Mazzone, S. B., Canning, B. J., and Weinreich, D. (2006). Reflex regulation of

airway sympathetic nerves in guinea-pigs. *J. Physiol.* 573, 549-564. doi:

10.1113/jphysiol.2005.104661.

O'Halloran, K. D., Curran, A. K., and Bradford, A. (1996). The effect of sympathetic nerve

stimulation on ventilation and upper airway resistance in the anaesthetized rat.

Adv. Exp. Med. Biol. 410, 443-447. doi: 10.1007/978-1-4615-5891-0_68

O'Halloran, K. D., Curran, A. K., and Bradford, A. (1998). Influence of cervical

sympathetic nerves on ventilation and upper airway resistance in the rat. *Eur.

Respir. J*. 12, 177–184. doi: 10.1183/09031936.98.12010177

Olson, E. B., Jr, Vidruk, E. H., and Dempsey, J. A. (1988). Carotid body excision

significantly changes ventilatory control in awake rats. *J. Appl. Physiol. (1985)*. 64,

666-671. doi: 10.1152/jappl.1988.64.2.666

O'Regan, R. G. (1981). Responses of carotid body chemosensory activity and blood flow

to stimulation of sympathetic nerves in the cat. *J Physiol*. 315, 81–98. doi:

10.1113/jphysiol.1981.sp013734

O'Regan, R. G. (1975). The influences exerted by the centrifugal innervation of the

carotid sinus nerv. In: Purves, MJ., editor. The Peripheral Arterial Chemoreceptors.

London: Cambridge University Press, 221–240.

Ortega-Sáenz, P., Levitsky, K. L., Marcos-Almaraz, M. T., Bonilla-Henao, V., Pascual, A.,

and López-Barneo, J. (2010). Carotid body chemosensory responses in mice

deficient of TASK channels. *J Gen Physiol*. 135, 379–392. doi:

10.1085/jgp.200910302

Ortega-Sáenz, P., Caballero, C., Gao, L., and López-Barneo, J. (2018). Testing Acute

Oxygen Sensing in Genetically Modified Mice: Plethysmography and Amperometry.

Methods Mol. Biol. 1742, 139-153. doi: 10.1007/978-1-4939-7665-2_13

Ortega-Sáenz, P., and López-Barneo, J. (2020). Physiology of the Carotid Body: From

Molecules to Disease. *Annu Rev Physiol*. 82, 127–149. doi: 10.1146/annurev-

physiol-020518-114427

Ortiz, F., Iturriaga, R., and Varas, R. (2009). Sustained hypoxia enhances TASK-like

current inhibition by acute hypoxia in rat carotid body type-I cells. *Adv Exp Med

Biol*. 648, 83–88. doi: 10.1007/978-90-481-2259-2_9

Overholt, J. L., and Prabhakar, N. R. (1999). Norepinephrine inhibits a toxin resistant Ca^{2+}

current in carotid body glomus cells: evidence for a direct G protein mechanism. *J.

Neurophysiol*. 81, 225–233. doi: 10.1152/jn.1999.81.1.225

Palmer, L. A., May, W. J., deRonde, K., Brown-Steinke, K., Bates, J. N., Gaston, B., and

Lewis, S. J. (2013a). Ventilatory responses during and following exposure to a

hypoxic challenge in conscious mice deficient or null in S-nitrosoglutathione

reductase. *Respir. Physiol. Neurobiol*. 185, 571-581. doi: 10.1016/j.resp.2012.11.009

Palmer, L. A., May, W. J., deRonde, K., Brown-Steinke, K., Gaston, B., and Lewis, S. J.

(2013b). Hypoxia-induced ventilatory responses in conscious mice: gender

differences in ventilatory roll-off and facilitation. *Respir. Physiol. Neurobiol*. 185,

497-505. doi: 10.1016/j.resp.2012.11.010

Palmer, L. A., Kimberly deRonde, Brown-Steinke, K., Gunter, S., Jyothikumar, V., Forbes,

M. S., and Lewis, S. J. (2015). Hypoxia-induced changes in protein s-nitrosylation in

female mouse brainstem. *Am. J. Respir. Cell Mol. Biol*. 52, 37-45. doi:

10.1165/rcmb.2013-0359OC

Pang, L., Miao, Z. H., Dong, L., and Wang, Y. L. (1999). *Sheng li xue bao : [Acta physiologica Sinica]*. 51, 407-412.

Pankevich, D. E., Deedy, E. M., Cherry, J. A., and Baum, M. J. (2003a). Interactive effects
of testosterone and superior cervical ganglionectomy on attraction thresholds to
volatile urinary odors in gonadectomized mice. *Behav. Brain Res*. 144, 157–165. doi:
10.1016/s0166-4328(03)00073-1

Pankevich, D., Baum, M. J., and Cherry, J. A. (2003b). Removal of the superior cervical
ganglia fails to block Fos induction in the accessory olfactory system of male mice
after exposure to female odors. *Neurosci. Lett*. 345, 13–16. doi: 10.1016/s0304-
3940(03)00471-3

Pardal, R., Ortega-Sáenz, P., Durán, R., and López-Barneo, J. (2007). Glia-like stem cells
sustain physiologic neurogenesis in the adult mammalian carotid body. *Cell*. 131,
364–377. doi: 10.1016/j.cell.2007.07.043

Pardal, R., Ludewig, U., Garcia-Hirschfeld, J., and Lopez-Barneo, J. (2000). Secretory
responses of intact glomus cells in thin slices of rat carotid body to hypoxia and
tetraethylammonium. *Proc Natl Acad Sci U S A*. 97, 2361–2366. doi:
10.1073/pnas.030522297

Pashai, P., Kostuk, E. W., Pilchard, L. E., and Shirahata, M. (2012). ATP release from the
carotid bodies of DBA/2J and A/J inbred mouse strains. *Adv. Exp. Med. Biol*. 758,
279–285. doi: 10.1007/978-94-007-4584-1_38

Paton, J. F., Sobotka, P. A., Fudim, M., Engelman, Z. J., Hart, E. C., McBryde, F. D., Abdala,
A. P., Marina, N., Gourine, A. V., Lobo, M., Patel, N., Burchell, A., Ratcliffe, L., and
Nightingale, A. (2013). The carotid body as a therapeutic target for the treatment of

sympathetically mediated diseases. *Hypertension*. 61, 5–13. doi:

10.1161/HYPERTENSIONAHA.111.00064

Patwardhan, R. V., Tubbs, R. S., Killingsworth, C. R., Rollins, D. L., Smith, W. M., and

Ideker, R. E. (2002). Ninth cranial nerve stimulation for epilepsy control. Part 1:

efficacy in an animal model. *Pediatr Neurosurg*. 36, 236–243. doi:

10.1159/000058426

Peers, C., and Green, F. K. (1991). Inhibition of Ca(2+)-activated K+ currents by

intracellular acidosis in isolated type I cells of the neonatal rat carotid body. *J

Physiol*. 437, 589–602. doi: 10.1113/jphysiol.1991.sp018613

Peers, C., and Buckler, K. J. (1995). Transduction of chemostimuli by the type I carotid

body cell. *J. Membr. Biol*. 144, 1–9. doi: 10.1007/bf00238411

Peng, Y. J., Overholt, J. L., Kline, D., Kumar, G. K., andPrabhakar, N. R. (2003). Induction

of sensory long-term facilitation in the carotid body by intermittent hypoxia:

implications for recurrent apneas. *Proc Natl Acad Sci U S A*. 100, 10073–10078. doi:

10.1073/pnas.1734109100

Peng, Y. J., Zhang, X., Nanduri, J., and Prabhakar, N. R. (2018). Therapeutic Targeting of

the Carotid Body for Treating Sleep Apnea in a Pre-clinical Mouse Model. *Adv. Exp.

Med. Biol*. 1071, 109–114. doi: 10.1007/978-3-319-91137-3_14

Pérez-García, M. T., Colinas, O., Miguel-Velado, E., Moreno-Domínguez, A., and López-

López, J. R. (2004). Characterization of the Kv channels of mouse carotid body

chemoreceptor cells and their role in oxygen sensing. *J. Physiol*. 557, 457–471. doi:

10.1113/jphysiol.2004.062281

Petheo, G. L., Molnár, Z., Róka, A., Makara, J. K., and Spät, A. (2001). A pH-sensitive

chloride current in the chemoreceptor cell of rat carotid body. *J Physiol*. 535, 95–

106. doi: 10.1111/j.1469-7793.2001.00095.x

Pichard, L. E., Crainiceanu, C. M., Pashai, P., Kostuk, E. W., Fujioka, A., and Shirahata, M.

(2015). Role of BK Channels in Murine Carotid Body Neural Responses in vivo. *Adv.*

Exp. Med. Biol. 860, 325-333. doi: 10.1007/978-3-319-18440-1_37

Pizarro, J., Warner, M. M., Ryan, M., Mitchell, G. S., and Bisgard, G. E. (1992).

Intracarotid norepinephrine infusions inhibit ventilation in goats. *Respir. Physiol*. 90,

299–310. doi: 10.1016/0034-5687(92)90110-i

Platero-Luengo, A., González-Granero, S., Durán, R., Díaz-Castro, B., Piruat, J. I., García-

Verdugo, J. M., Pardal, R., and López-Barneo, J. (2014). An O2-sensitive glomus cell-

stem cell synapse induces carotid body growth in chronic hypoxia. *Cell*, 156, 291–

303. doi: 10.1016/j.cell.2013.12.013

Pokorski, M., and Antosiewicz, J. (2010). Alterations in the hypoxic ventilatory response

with advancing age in awake rats. *J. Physiol. Pharmacol*. 61, 227–232.

Poncet, L., Denoroy, L., Dalmaz, Y., Péquignot, J. M., and Jouvet, M. (1994). Chronic

hypoxia affects peripheral and central vasoactive intestinal peptide-like

immunoreactivity in the rat. *Neurosci Lett*. 176, 1–4. doi: 10.1016/0304-

3940(94)90856-7

Porzionato, A., Macchi, V., Parenti, A., and De Caro, R. (2008). Trophic factors in the

carotid body. *Int Rev Cell Mol Biol*. 269, 1–58. doi: 10.1016/S1937-6448(08)01001-0

Porzionato, A., Macchi, V., Stecco, C., and De Caro, R. (2019). The Carotid Sinus Nerve-Structure, Function, and Clinical Implications. *Anat Rec (Hoboken)*. 302, 575–587. doi: 10.1002/ar.23829

Potter, E. K., and McCloskey, D. I. (1987). Excitation of carotid body chemoreceptors by neuropeptide-Y. *Respir. Physiol.* 67, 357–365. doi: 10.1016/0034-5687(87)90065-x

Powell, F. L., Milsom, W. K., and Mitchell, G. S. (1998). Time domains of the hypoxic ventilatory response. *Respir. Physiol.* 112, 123–134. doi: 10.1016/s0034-5687(98)00026-7

Powell F. L. (2007). The influence of chronic hypoxia upon chemoreception. *Respir Physiol Neurobiol.* 157, 154–161. doi: 10.1016/j.resp.2007.01.009

Prabhakar, N. R., Kou, Y. R., Cragg, P. A., and Cherniack, N. S. (1993). Effect of arterial chemoreceptor stimulation: role of norepinephrine in hypoxic chemotransmission. *Adv. Exp. Med. Biol.* 337, 301-306. doi: 10.1007/978-1-4615-2966-8_42

Prabhakar, N. R. (1994). Neurotransmitters in the carotid body. *Adv. Exp. Med. Biol.* 360, 57–69. doi: 10.1007/978-1-4615-2572-1_6

Prabhakar, N. R., Peng, Y. J., and Nanduri, J. (2018). Recent advances in understanding the physiology of hypoxic sensing by the carotid body. *F1000Res.* 7, F1000 Faculty Rev-1900. doi: 10.12688/f1000research.16247.1

Prabhakar N. R. (2000). Oxygen sensing by the carotid body chemoreceptors. *J. Appl. Physiol. (1985).* 88, 2287–2295. doi: 10.1152/jappl.2000.88.6.2287

Prabhakar, N. R. and Overholt, J. L. (2000). Cellular mechanisms of oxygen sensing at the carotid body: heme proteins and ions channels. *Respir Physiol.* 122, 209–221. doi:

10.1016/S0034-5687(00)00160-2

Price, E. R., Han, F., Dick, T. E., and Strohl, K. P. (2003). 7-nitroindazole and posthypoxic

ventilatory behavior in the A/J and C57BL/6J mouse strains. *J. Appl. Physiol. (1985)*.

95, 1097–1104. doi: 10.1152/japplphysiol.00166.2003

Prieto-Lloret, J., Donnelly, D. F., Rico, A. J., Moratalla, R., González, C., and Rigual, R. J.

(2007). Hypoxia transduction by carotid body chemoreceptors in mice lacking

dopamine D(2) receptors. *J. Appl. Physiol. (1985)*. 103, 1269–1275. doi:

10.1152/japplphysiol.00391.2007

Purves M. J. (1970). The role of the cervical sympathetic nerve in the regulation of

oxygen consumption of the carotid body of the cat. *J Physiol*. 209, 417–431. doi:

10.1113/jphysiol.1970.sp009172

Putnam, R. W., Conrad, S. C., Gdovin, M. J., Erlichman, J. S., and Leiter, J. C. (2005).

Neonatal maturation of the hypercapnic ventilatory response and central neural

CO_2 chemosensitivity. *Respir. Physiol. Neurobiol*. 149, 165-179. doi:

10.1016/j.resp.2005.03.004

Qu, L., Sherebrin, R., and Weaver, L. C. (1988). Blockade of spinal pathways decreases

pre- and postganglionic discharge differentially. *Am. J. Physiol*. 255, R946-R951. doi:

10.1152/ajpregu.1988.255.6.R946

Rando, T. A., Bowers, C. W., and Zigmond, R. E. (1981). Localization of neurons in the rat

spinal cord which project to the superior cervical ganglion. *J. Comp. Neurol*. 196, 73-

83. doi: 10.1002/cne.901960107

Rees, P. M. (1967). Observations on the fine structure and distribution of presumptive baroreceptor nerves at the carotid sinus. *J. Comp. Neurol.* 131, 517–548. doi: 10.1002/cne.901310409

Rey, S., Del Rio, R., Alcayaga, J., and Iturriaga, R. (2004). Chronic intermittent hypoxia enhances cat chemosensory and ventilatory responses to hypoxia. *J Physiol.* 560, 577–586. doi: 10.1113/jphysiol.2004.072033

Rigual, R., Cachero, M. T., Rocher, A., and González, C. (1999). Hypoxia inhibits the synthesis of phosphoinositides in the rabbit carotid body. *Pflugers Arch.* 437, 839-845. doi: 10.1007/s004240050853

Rivas-Ramírez, P., Reboreda, A., Rueda-Ruzafa, L., Herrera-Pérez, S., and Lamas, J. A. (2020). PIP_2 Mediated Inhibition of TREK Potassium Currents by Bradykinin in Mouse Sympathetic Neurons. *Int. J. Mol. Sci.* 21, 389. doi: 10.3390/ijms21020389

Rodenbaugh, D. W., Collins, H. L., Nowacek, D. G., and DiCarlo, S. E. (2003). Increased susceptibility to ventricular arrhythmias is associated with changes in Ca^{2+} regulatory proteins in paraplegic rats. *Am. J. Physiol. Heart. Circ. Physiol.* 285, H2605–H2613. doi: 10.1152/ajpheart.00319.2003

Rong, W., Gourine, A. V., Cockayne, D. A., Xiang, Z., Ford, A. P., Spyer, K. M., and Burnstock, G. (2003). Pivotal role of nucleotide P2X2 receptor subunit of the ATP-gated ion channel mediating ventilatory responses to hypoxia. *J Neurosci.* 23, 11315–11321. doi: 10.1523/JNEUROSCI.23-36-11315.2003

Roux, J. C., Dura, E., and Villard, L. (2008). Tyrosine hydroxylase deficit in the chemoafferent and the sympathoadrenergic pathways of the Mecp2 deficient mouse. *Neurosci. Lett.* 447, 82–86. doi: 10.1016/j.neulet.2008.09.045

Roux, J. C., Peyronnet, J., Pascual, O., Dalmaz, Y., and Pequignot, J. M. (2000).

Ventilatory and central neurochemical reorganisation of O_2 chemoreflex after

carotid sinus nerve transection in rat. *J. Physiol.* 522 Pt 3, 493–501. doi:

10.1111/j.1469-7793.2000.t01-4-00493.x

Ryan, M. L., Hedrick, M. S., Pizarro, J., and Bisgard, G. E. (1995). Effects of carotid body

sympathetic denervation on ventilatory acclimatization to hypoxia in the goat.

Respir. Physiol. 99, 215-224. doi: 10.1016/0034-5687(94)00096-i

Saavedra, J. M. (1985). Central and peripheral catecholamine innervation of the rat

intermediate and posterior pituitary lobes. *Neuroendocrinology.* 40, 281-284. doi:

10.1159/000124087

Sadoshima, S., Busija, D., Brody, M., and Heistad, D. (1981). Sympathetic nerves protect

against stroke in stroke-prone hypertensive rats. A preliminary

report. *Hypertension.* 3, I124-I127. doi: 10.1161/01.hyp.3.3_pt_2.i124

Sadoshima, S., and Heistad, D. D. (1983). Regional cerebral blood flow during

hypotension in normotensive and stroke-prone spontaneously hypertensive rats:

effect of sympathetic denervation. *Stroke.* 14, 575-579. doi:

10.1161/01.str.14.4.575

Sankari, A., Bascom, A., Oomman, S., & Badr, M. S. (2014). Sleep disordered breathing in

chronic spinal cord injury. *J. Clin. Sleep Med.* 10, 65–72. doi: 10.5664/jcsm.3362

Sankari, A., Badr, M. S., Martin, J. L., Ayas, N. T., and Berlowitz, D. J. (2019). Impact Of

Spinal Cord Injury On Sleep: Current Perspectives. *Nat. Sci. Sleep.* 11, 219-229. doi:

10.2147/NSS.S197375

Santos, M. S., Moreno, A. J., and Carvalho, A. P. (1996). Relationships between ATP depletion, membrane potential, and the release of neurotransmitters in rat nerve terminals. An in vitro study under conditions that mimic anoxia, hypoglycemia, and ischemia. *Stroke.* 27, 941-950. doi: 10.1161/01.str.27.5.941

Sapru, H. N., and Krieger, A. J. (1977). Carotid and aortic chemoreceptor function in the rat. *J. Appl. Physiol. Respir. Environ. Exerc. Physiol.* 42, 344–348. doi: 10.1152/jappl.1977.42.3.344

Savastano, L. E., Castro, A. E., Fitt, M. R., Rath, M. F., Romeo, H. E., and Muñoz, E. M. (2010). A standardized surgical technique for rat superior cervical ganglionectomy. *J. Neurosci. Methods.* 192, 22–33. doi: 10.1016/j.jneumeth.2010.07.007

Schlaefke, M. E., Kille, J. F., and Loeschcke, H. H. (1979). Elimination of central chemosensitivity by coagulation of a bilateral area on the ventral medullary surface in awake cats. *Pflugers Arch.* 378, 231–241. doi: 10.1007/BF00592741

Schömig, A., Fischer, S., Kurz, T., Richardt, G., and Schömig, E. (1987). Nonexocytotic release of endogenous noradrenaline in the ischemic and anoxic rat heart: mechanism and metabolic requirements. *Circ. Res.* 60, 194–205. doi: 10.1161/01.res.60.2.194

Sengupta, P. (2013). The Laboratory Rat: Relating Its Age With Human's. *Int. J. Prev. Med.* 4, 624-630.

Serra, A., Brozoski, D., Hedin, N., Franciosi, R., and Forster, H. V. (2001). Mortality after carotid body denervation in rats. J. Appl. Physiol. (1985). 91, 1298–1306. doi: 10.1152/jappl.2001.91.3.1298

Serra, A., Brozoski, D., Hodges, M., Roethle, S., Franciosi, R., and Forster, H. V. (2002).
Effects of carotid and aortic chemoreceptor denervation in newborn piglets. J. Appl.
Physiol. (1985). 92, 893–900. doi: 10.1152/japplphysiol.00819.2001

Sheikhbahaei, S., Turovsky, E. A., Hosford, P. S., Hadjihambi, A., Theparambil, S. M., Liu,
B., Marina, N., Teschemacher, A. G., Kasparov, S., Smith, J. C., and Gourine, A. V.
(2016). Astrocytes modulate brainstem respiratory rhythm-generating circuits and
determine exercise capacity. *Nat. Commun.* 9, 370. doi: 10.1038/s41467-017-
02723-6.

Shin, J. C., Han, E. Y., Cho, K. H., and Im, S. H. (2019). Improvement in Pulmonary
Function with Short-term Rehabilitation Treatment in Spinal Cord Injury Patients.
Sci. Rep. 9, 17091. doi: 10.1038/s41598-019-52526-6

Simeone, X., Karch, R., Ciuraszkiewicz, A., Orr-Urtreger, A., Lemmens-Gruber, R.,
Scholze, P., and Huck, S. (2019). The role of the nAChR subunits 5, 2, àȁ nd 4 onˋ
synaptic transmission in the mouse superior cervical ganglion. *Physiol. Rep.* 7,
e14023. doi: 10.14814/phy2.14023

Sinclair, J. D. (1987). Respiratory drive in hypoxia: carotid body and other mechanisms
compared. *News Pharmacol. Sci.* 2, 57–60. doi:
10.1152/physiologyonline.1987.2.2.57

Sinclair, D., Purves-Tyson, T. D., Allen, K.M., and Weickert, C. S. (2014). Impacts of stress
and sex hormones on dopamine neurotransmission in the adolescent brain.
Psychopharmacology (Berl). 231, 1581-1599. doi: 10.1007/s00213-013-3415-z.

Smith, P. G., and Mills, E. (1980). Restoration of reflex ventilatory response to hypoxia

after removal of carotid bodies in the cat. *Neuroscience*. 5, 573–580. doi:

10.1016/0306-4522(80)90054-8

Smith, C. A., Jameson, L. C., Mitchell, G. S., Musch, T. I., and Dempsey, J. A. (1984).

Central-peripheral chemoreceptor interaction in awake cerebrospinal fluid-

perfused goats. *J. Appl. Physiol.* 56, 1541–1549. doi:10.1152/jappl.1984.56.6.1541

Smith, C. A., Rodman, J. R., Chenuel, B. J., Henderson, K. S., and Dempsey, J. A. (2006).

Response time and sensitivity of the ventilatory response to CO2 in unanesthetized

intact dogs: central vs. peripheral chemoreceptors. *J. Appl. Physiol. (1985)*. 100, 13–

19. doi: 10.1152/japplphysiol.00926.2005

Smith, C. A., Forster, H. V., Blain, G. M., and Dempsey, J. A. (2010). An interdependent

model of central/peripheral chemoreception: Evidence and implications for

ventilatory control. *Respir. Physiol. Neurobiol.* 173, 288–297. doi:

10.1016/j.resp.2010.02.015

Smith, C. A., Blain, G. M., Henderson, K. S., and Dempsey, J. A. (2015). Peripheral

chemoreceptors determine the respiratory sensitivity of central chemoreceptors to

CO_2: role of carotid body CO_2. *J. Physiol.* 593, 4225–4243. doi: 10.1113/JP270114

Smith, C. A., Nakayama, H., and Dempsey, J. A. (2003). The essential role of carotid body

chemoreceptors in sleep apnea. *Can. J. Physiol. Pharmacol.* 81, 774-779. doi:

10.1139/y03-056

Sobrino, V., González-Rodríguez, P., Annese, V., López-Barneo, J., and Pardal, R. (2018).

Fast neurogenesis from carotid body quiescent neuroblasts accelerates adaptation to hypoxia. *EMBO reports.* 19, e44598. doi: 10.15252/embr.201744598

Solberg, L. C., Valdar, W., Gauguier, D., Nunez, G., Taylor, A., Burnett, S., Arboledas-Hita, C., Hernandez-Pliego, P., Davidson, S., Burns, P., Bhattacharya, S., Hough, T., Higgs, D., Klenerman, P., Cookson, W. O., Zhang, Y., Deacon, R. M., Rawlins, J. N., Mott, R., and Flint, J. (2006). A protocol for high-throughput phenotyping, suitable for quantitative trait analysis in mice. *Mamm. Genome.* 17, 129-146. doi: 10.1007/s00335-005-0112-1

Song, G., Xu H., Wang, H., Macdonald, S. M., Poon, C-S. (2011). Hypoxia-excited neurons in NTS send axonal projections to Kölliker-Fuse/parabrachial complex in dorsolateral pons. *Neuroscience.* 175, 145-153. doi: 10.1016/j.neuroscience.2010.11.06.

Soulier, V., Gestreau, C., Borghini, N., Dalmaz, Y., Cottet-Emard, J.M., and Pequignot, J.M. (1997). Peripheral chemosensitivity and central integration: neuroplasticity of catecholaminergic cells under hypoxia. *Comp. Biochem. Physiol. A. Physiol.* 118, 1-7. doi: 10.1016/s0300-9629(96)00369-6

Stein, R. D., and Weaver, L. C. (1988). Multi- and single-fibre mesenteric and renal sympathetic responses to chemical stimulation of intestinal receptors in cats. *J. Physiol.* 396, 155-172. doi: 10.1113/jphysiol.1988.sp016956

Strohl, K. P. (2003). Periodic breathing and genetics. *Respir. Physiol. Neurobiol.* 135, 179-185. doi: 10.1016/s1569-9048(03)00036-3

Strosznajder, R. P. (1997). Effect of hypoxia and dopamine on arachidonic acid metabolism in superior cervical ganglion. *Neurochem. Res.* 22, 1193–1197. doi:

10.1023/a:1021920610766

Stunden, C. E., Filosa, J. A., Garcia, A. J., Dean, J. B., and Putnam, R. W. (2001). Development of in vivo ventilatory and single chemosensitive neuron responses to hypercapnia in rats. *Respir. Physiol.* 127, 135–155. doi: 10.1016/s0034-5687(01)00242-0

Suguihara, C., Hehre, D., and Bancalari, E. (1994). Effect of dopamine on hypoxic ventilatory response of sedated piglets with intact and denervated carotid bodies. *J. Appl. Physiol. (1985).* 77, 285–289. doi: 10.1152/jappl.1994.77.1.285

Summers, B. A., Overholt, J. L., and Prabhakar, N. R. (2002). CO(2) and pH independently modulate L-type Ca(2+) current in rabbit carotid body glomus cells. *J Neurophysiol.* 88, 604–612. doi: 10.1152/jn.2002.88.2.604

Sun, M. K., and Reis, D. J. (1994). Hypoxia selectively excites vasomotor neurons of rostral ventrolateral medulla in rats. *Am. J. Physiol.* 266, R245–R256. doi: 10.1152/ajpregu.1994.266.1.R245

Tagaito, Y., Polotsky, V. Y., Campen, M. J., Wilson, J. A., Balbir, A., Smith, P. L., Schwartz, A. R., and O'Donnell, C. P. (2001). A model of sleep-disordered breathing in the C57BL/6J mouse. *J. Appl. Physiol. (1985).* 91, 2758–2766. doi: 10.1152/jappl.2001.91.6.2758

Tahayori, B., and Koceja, D. M. (2012). Activity-dependent plasticity of spinal circuits in the developing and mature spinal cord. *Neural Plast.* 2012, 964843. doi: 10.1155/2012/964843.

Takahashi, M., Matsuda, H., Hayashida, Y., Yamamoto, Y., Tsukuda, M., and Kusakabe, T. (2011). Morphological characteristics and peptidergic innervation in the carotid body of spontaneously hypertensive rats. *Histol Histopathol*. 26, 369–375. doi: 10.14670/HH-26.369

Takaki, F., Nakamuta, N., Kusakabe, T., and Yamamoto, Y. (2015). Sympathetic and sensory innervation of small intensely fluorescent (SIF) cells in rat superior cervical ganglion. *Cell Tissue Res*. 359, 441-451. doi: 10.1007/s00441-014-2051-1.

Tan, Z. Y., Lu, Y., Whiteis, C. A., Benson, C. J., Chapleau, M. W., and Abboud, F. M. (2007). Acid-sensing ion channels contribute to transduction of extracellular acidosis in rat carotid body glomus cells. *Circ Res*. 101, 1009–1019. doi: 10.1161/CIRCRESAHA.107.154377

Tang, F. R., Tan, C. K., and Ling, E. A. (1995a). An ultrastructural study of the sympathetic preganglionic neurons that innervate the superior cervical ganglion in spontaneously hypertensive rats and Wistar-Kyoto rats. *J. Hirnforsch*. 36, 411–420.

Tang, F. R., Tan, C. K., and Ling, E. A. (1995b). A comparative study by retrograde neuronal tracing and substance P immunohistochemistry of sympathetic preganglionic neurons in spontaneously hypertensive rats and Wistar-Kyoto rats. *J. Anat*. 186, 197–207.

Tang, F. R., Tan, C. K., and Ling, E. A. (1995c). A comparative study of NADPH-diaphorase in the sympathetic preganglionic neurons of the upper thoracic cord between spontaneously hypertensive rats and Wistar-Kyoto rats. *Brain Res*. 691, 153–159. doi: 10.1016/0006-8993(95)00658-d

Tankersley, C. G., Fitzgerald, R. S., and Kleeberger, S. R. (1994). Differential control of ventilation among inbred strains of mice. *Am. J. Physiol.* 267, R1371–R1377. doi: 10.1152/ajpregu.1994.267.5.R1371

Tankersley, C. G., Elston, R. C., and Schnell, A. H. (2000). Genetic determinants of acute hypoxic ventilation: patterns of inheritance in mice. *J. Appl. Physiol. (1985).* 88, 2310–2318. doi: 10.1152/jappl.2000.88.6.2310

Tankersley, C. G. (2001). Selected contribution: variation in acute hypoxic ventilatory response is linked to mouse chromosome 9. *J. Appl. Physiol. (1985).* 90, 1615–1606. doi: 10.1152/jappl.2001.90.4.1615

Tankersley, C. G., Irizarry, R., Flanders, S., and Rabold, R. (2002). Circadian rhythm variation in activity, body temperature, and heart rate between C3H/HeJ and C57BL/6J inbred strains. *J. Appl. Physiol. (1985).* 92, 870–877. doi: 10.1152/japplphysiol.00904.2001

Tankersley, C. G. (2003). Genetic aspects of breathing: on interactions between hypercapnia and hypoxia. *Respir. Physiol. Neurobiol.* 135, 167–178. doi: 10.1016/s1569-9048(03)00035-1

Teppema, L. J., and Dahan, A. (2010). The ventilatory response to hypoxia in mammals: mechanisms, measurement, and analysis. *Physiol. Rev.* 90, 675-754. doi: 10.1152/physrev.00012.2009

Teshima, T., Tucker, A. S., and Lourenço, S. V. (2019). Dual Sympathetic Input into Developing Salivary Glands. *J. Dent. Res.* 98, 1122–1130. doi: 10.1177/0022034519865222

Tester, N. J., Fuller, D. D., Fromm, J. S., Spiess, M. R., Behrman, A. L., and Mateika, J. H. (2014). Long-term facilitation of ventilation in humans with chronic spinal cord injury. *Am. J. Respir. Crit. Care Med.* 189, 57–65. doi: 10.1164/rccm.201305-0848OC

Tewari, S. G., Bugenhagen, S. M., Wang, Z., Schreier, D. A., Carlson, B. E., Chesler, N. C., and Beard, D. A. (2013). Analysis of cardiovascular dynamics in pulmonary hypertensive C57BL6/J mice. *Front. Physiol.* 4, 355. doi: 10.3389/fphys.2013.00355

Thannickal, T. C., Moore, R. Y., Nienhuis, R., Ramanathan, L., Gulyani, S., Aldrich, M., Cornford, M., and Siegel, J. M. (2000). Reduced number of hypocretin neurons in human narcolepsy *Neuron.* 27, 469–474. doi: 10.1016/s0896-6273(00)00058-1

Timmers, H. J., Wieling, W., Karemaker, J. M., and Lenders, J. W. (2003). Denervation of carotid baro- and chemoreceptors in humans. *J Physiol.* 553, 3–11. doi: 10.1113/jphysiol.2003.052415

Torrealba, F., and Claps, A. (1988). The carotid sinus connections: a WGA-HRP study in the cat. *Brain Res.* 455, 134–143. doi: 10.1016/0006-8993(88)90122-9

Tosaka, T., and Kobayashi, H. (1977). The SIF cell as a functional modulator of ganglionic transmission through the release of dopamine. Arch. Histol. Jpn. 40 Suppl, 187-196. doi: 10.1679/aohc1950.40.supplement_187

Tse, A., Yan, L., Lee, A. K., and Tse, F. W. (2012). Autocrine and paracrine actions of ATP in rat carotid body. *Can. J. Physiol. Pharmacol.* 90, 705–711. doi: 10.1139/y2012-054

Tubbs, R. S., Patwardhan, R. V., and Oakes, W. J. (2002). Ninth cranial nerve stimulation

for epilepsy control. Part 2: surgical feasibility in humans. *Pediatr Neurosurg.* 36,

244–247. doi: 10.1159/000058427

Turovsky, E., Theparambil, S. M., Kasymov, V., Deitmer, J. W., Del Arroyo, A. G., Ackland

G. L., Corneveaux J. J., Allen, A.N., Huentelman, M. J., Kasparov, S., Marina, N., and

Gourine, A. V. (2016). Mechanisms of CO_2/H^+ Sensitivity of Astrocytes. *J. Neurosci.*

36, 10750-10758. doi: 10.1523/JNEUROSCI.1281-16.2016

Ureña, J., Fernández-Chacón, R., Benot, A. R., Alvarez de Toledo, G. A., and López-

Barneo, J. (1994). Hypoxia induces voltage-dependent Ca^{2+} entry and quantal

dopamine secretion in carotid body glomus cells. *Proc Natl Acad Sci U S A.* 91,

10208–10211. doi: 10.1073/pnas.91.21.10208

Vázquez-Nin, G. H., Costero, I, Echeverría, O. M., Aguilar, R., and Barroso-Moguel, R.

(1978). Innervation of the carotid body. An experimental quantitative study. *Acta

Anat (Basel).* 102, 12-28. doi: 10.1159/000145613

Verna, A., Barets, A., and Salat, C. (1984). Distribution of sympathetic nerve endings

within the rabbit carotid body: a histochemical and ultrastructural study. *J.

Neurocytol.* 13, 849-865. doi: 10.1007/BF01148589

Villiere, S. M., Nakase, K., Kollmar, R., Silverman, J., Sundaram, K., and Stewart, M.

(2017). Seizure-associated central apnea in a rat model: Evidence for resetting the

respiratory rhythm and activation of the diving reflex. *Neurobiol. Dis.* 101, 8–15.

doi: 10.1016/j.nbd.2017.01.008

Vizek, M., Pickett, C. K., and Weil, J. V. (1987). Biphasic ventilatory response of adult cats to sustained hypoxia has central origin. *J. Appl. Physiol. (1985)*. 63, 1658-1664. doi: 10.1152/jappl.1987.63.4.1658

Wagenaar, M., Teppema, L., Berkenbosch, A., Olievier, C., and Folgering, H. (1998). Effect of low-dose acetazolamide on the ventilatory CO2 response during hypoxia in the anaesthetized cat. *Eur Respir J*. 12, 1271–1277. doi: 10.1183/09031936.98.12061271

Wakai, J., Takamura, D., Morinaga, R., Nakamuta, N., and Yamamoto, Y. (2015). Differences in respiratory changes and Fos expression in the ventrolateral medulla of rats exposed to hypoxia, hypercapnia, and hypercapnic hypoxia. *Respir. Physiol. Neurobiol*. 215, 64–72. doi: 10.1016/j.resp.2015.05.008

Wallenstein, S., Zucker, C. L., and Fleiss, J. L. (1980). Some statistical methods useful in circulation research. *Circ. Res*. 47, 1-9. doi: 10.1161/01.res.47.1.1

Wang, Z. Y., and Bisgard, G. E. (2005). Postnatal growth of the carotid body. *Respir Physiol Neurobiol*. 149, 181–190. doi: 10.1016/j.resp.2005.03.016

Wang, H. W., and Chiou, W. Y. (2004). Sympathetic innervation of the tongue in rats *ORL J. Otorhinolaryngol. Relat. Spec*. 66, 16–20. doi: 10.1159/000077228

Wang, J., Hogan, J. O., Wang, R., White, C., and Kim, D. (2017). Role of cystathionine-γ-lyase in hypoxia-induced changes in TASK activity, intracellular $[Ca^{2+}]$ and ventilation in mice. *Respir. Physiol. Neurobiol*. 246, 98–106. doi: 10.1016/j.resp.2017.08.009

Weir, E. K., López-Barneo, J., Buckler, K. J., and Archer, S. L. (2005). Acute oxygen-sensing mechanisms. *N Engl J Med*. 353, 2042–2055. doi: 10.1056/NEJMra050002

Werber, A. H., and Heistad, D. D. (1984). Effects of chronic hypertension and sympathetic nerves on the cerebral microvasculature of stroke-prone

spontaneously hypertensive rats. *Circ. Res.* 55, 286-294. doi:

10.1161/01.res.55.3.286

Westerhaus, M. J., and Loewy, A. D. (1999). Sympathetic-related neurons in the preoptic

region of the rat identified by viral transneuronal labeling. *J. Comp. Neurol.* 414,

361-378.

Wiberg, M., and Widenfalk, B. (1993). Involvement of connections between the

brainstem and the sympathetic ganglia in the pathogenesis of rheumatoid arthritis.

An anatomical study in rats. *Scand. J. Plast. Reconstr. Surg. Hand Surg.* 27, 269-276.

Wilkinson, M. H., Berger, P. J., Blanch, N., Brodecky, V., and Jones, C. A. (1997).

Paradoxical effect of oxygen administration on breathing stability following post-

hyperventilation apnoea in lambs. *J. Physiol.* 504, 199-209. doi: 10.1111/j.1469-

7793.1997.199bf.x

Xie, A., Skatrud, J. B., Khayat, R., Dempsey, J. A., Morgan, B., and Russell, D. (2005).

Cerebrovascular response to carbon dioxide in patients with congestive heart

failure. *Am. J. Respir. Crit. Care Med.* 172, 371–378. doi: 10.1164/rccm.200406-

807OC

Xu, J., Xu, F., Tse, F. W., and Tse, A. (2005). ATP inhibits the hypoxia response in type I

cells of rat carotid bodies. *J Neurochem.* 92, 1419–1430. doi: 10.1111/j.1471-

4159.2004.02978.x

Yamaguchi, S., Balbir, A., Schofield, B., Coram, J., Tankersley, C. G., Fitzgerald, R. S.,

O'Donnell, C. P., and Shirahata, M. (2003). Structural and functional differences of the carotid body between DBA/2J and A/J strains of mice. *J. Appl. Physiol (1985)*. 94,

1536-1542. doi: 10.1152/japplphysiol.00739.2002

Yamaguchi, S., Balbir, A., Okumura, M., Schofield, B., Coram, J., Tankersley, C. G.,

Fitzgerald, R. S., O'Donnell, C. P., and Shirahata, M. (2006). Genetic influence on

carotid body structure in DBA/2J and A/J strains of mice. *Adv. Exp. Med. Biol*. 580,

105-359. doi: 10.1007/0-387-31311-7_16

Yamauchi, M., Dostal, J., Kimura, H., and Strohl, K. P. (2008a). Effects of buspirone on

posthypoxic ventilatory behavior in the C57BL/6J and A/J mouse strains. *J. Appl.

Physiol. (1985)*. 105, 518–526. doi: 10.1152/japplphysiol.00069.2008

Yamauchi, M., Ocak, H., Dostal, J., Jacono, F. J., Loparo, K. A., and Strohl, K. P. (2008b).

Post-sigh breathing behavior and spontaneous pauses in the C57BL/6J (B6) mouse.

Respir. Physiol. Neurobiol. 162, 117–125. doi: 10.1016/j.resp.2008.05.003

Yamauchi, M., Dostal, J., and Strohl, K. P. (2008c). Post-hypoxic unstable breathing in the

C57BL/6J mouse: effects of acetazolamide. *Adv. Exp. Med. Biol*. 605, 75-79. doi:

10.1007/978-0-387-73693-8_13

Yamauchi, M., Kimura, H., and Strohl, K. P. (2010). Mouse models of apnea: strain

differences in apnea expression and its pharmacologic and genetic

modification. *Adv. Exp. Med. Biol*. 669, 303–307. doi: 10.1007/978-1-4419-5692-

7_62

Yokota, R., and Yamauchi, A. (1974). Ultrastructure of the mouse superior cervical

ganglion, with particular reference to the pre- and postganglionic elements covering

the soma of its principal neurons. *Am. J. Anat*. 140, 281-297. doi:

10.1002/aja.1001400211

Yokoyama, T., Nakamuta, N., Kusakabe, T., and Yamamoto, Y. (2015). Sympathetic regulation of vascular tone via noradrenaline and serotonin in the rat carotid body as revealed by intracellular calcium imaging. *Brain Res*. 1596, 126-135. doi: 10.1016/j.brainres.2014.11.037

Young, A. P., Gruber, R. B., Discala, J. F., May, W. J., McLaughlin, D., Palmer, L. A., and Lewis, S. J. (2013). Co-activation of μ- and δ-opioid receptors elicits tolerance to morphine-induced ventilatory depression via generation of peroxynitrite. *Respir. Physiol. Neurobiol*. 186, 255–264. doi: 10.1016/j.resp.2013.02.028

Yuan, G., Peng, Y. J., Reddy, V. D., Makarenko, V. V., Nanduri, J., Khan, S. A., Garcia, J. A., Kumar, G. K., Semenza, G. L., and Prabhakar, N. R. (2013). Mutual antagonism between hypoxia-inducible factors 1α and 2α regulates oxygen sensing and cardio-respiratory homeostasis. *Proc Natl Acad Sci U S A*. 110, E1788–E1796. doi: 10.1073/pnas.1305961110

Zaidi, Z.F., and Matthews, M.R. (2013). Source and origin of nerve fibres immunoreactive for substance P and calcitonin gene-related peptide in the normal and chronically denervated superior cervical sympathetic ganglion of the rat. *Auton. Neurosci*. 173, 28-38. doi: 10.1016/j.autneu.2012.11.002

Zapata, P., Hess, A., Bliss, E. L., and Eyzaguirre, C. (1969). Chemical, electron microscopic and physiological observations on the role of catecholamines in the carotid body. *Brain Res*. 14, 473-496. doi: 10.1016/0006-8993(69)90123-1

Zapata, P. (1975). Effects of dopamine on carotid chemo- and baroreceptors in vitro. *J.*

Physiol. 244, 235-251. doi: 10.1113/jphysiol.1975.sp010794

Zhang, M., Zhong, H., Vollmer, C., and Nurse, C. A. (2000). Co-release of ATP and ACh mediates hypoxic signalling at rat carotid body chemoreceptors. *J Physiol.* 525 Pt 1, 143–158. doi: 10.1111/j.1469-7793.2000.t01-1-00143.x

Zhang, M., and Nurse, C. A. (2004). CO2/pH chemosensory signaling in co-cultures of rat carotid body receptors and petrosal neurons: role of ATP and ACh. *J Neurophysiol.* 92, 3433–3445. doi: 10.1152/jn.01099.2003

Zhang, M., Vollmer, C., and Nurse, C. A. (2018). Adenosine and dopamine oppositely modulate a hyperpolarization-activated current I_h in chemosensory neurons of the rat carotid body in co-culture. *J Physiol.* 596, 3101–3117. doi: 10.1113/JP274743

Ziegler, K. A., Ahles, A., Wille, T., Kerler, J., Ramanujam, D., and Engelhardt, S. (2018). Local sympathetic denervation attenuates myocardial inflammation and improves cardiac function after myocardial infarction in mice. *Cardiovasc. Res.* 114, 291-299. doi: 10.1093/cvr/cvx227